Characterization of Radiation Damage by Transmission Electron Microscopy

Series in Microscopy in Materials Science

Series Editors: **B Cantor**
M J Goringe

Other titles in the series

Atlas of Backscattering Kikuchi Diffraction Patterns
D J Dingley, K Z Baba-Kishi and V Randle
ISBN: 0 7503 0212 7

Electron Microscopy of Interfaces in Metals and Alloys
C T Forwood and L M Clarebrough
ISBN: 0 7503 0116 3

The Measurement of Grain Boundary Geometry
V Randle
ISBN: 0 7503 0235 6

Topics in Electron Diffraction and Microscopy of Materials
P Hirsch (ed)
ISBN: 0 7503 0538 X

Forthcoming titles in the series

Convergent Beam Electron Diffraction
P A Midgley and M Saunders

Electron Microscopy of Quasicrystals
K Chattopadhyay and S Ranganathan

Electron Microscopy of Catalysts
P Gai-Boyes

Forthcoming topics in the series

Magnetic Materials
Microscopy of Ceramics
Microscopy and Microanalysis
Orientation Imaging Microscopy
Phase Transformations
STEM
Semiconductor Materials

Series in Microscopy in Materials Science

Characterization of Radiation Damage by Transmission Electron Microscopy

M L Jenkins

Department of Materials
University of Oxford, UK

M A Kirk

Materials Science Division
Argonne National Laboratory, USA

Institute of Physics Publishing
Bristol and Philadelphia

British Library Cataloguing-in-Publication Data

A catalogue record for this book is available from the British Library.

ISBN 0 7503 0748 X (hbk)

Library of Congress Cataloging-in-Publication Data are available

Series Editors: **B Cantor**
M J Goringe

Commissioning Editor: Sally Wride
Production Editor: Simon Laurenson
Production Control: Sarah Plenty
Cover Design: Victoria Le Billon
Marketing Executive: Colin Fenton

Published by Institute of Physics Publishing, wholly owned by The Institute of Physics, London

Institute of Physics Publishing, Dirac House, Temple Back, Bristol BS1 6BE, UK

US Office: Institute of Physics Publishing, The Public Ledger Building, Suite 1035, 150 South Independence Mall West, Philadelphia, PA 19106, USA

Typeset in TEX using the IOP Bookmaker Macros
Printed in the UK by MPG Books Ltd, Bodmin

Contents

We wish to acknowledge the following for permission to reproduce figures.

Academic Press: Brown C 1976 *Proceedings of EMAG 75* ed J A Venebles, p 405. American Institute of Physics: Allen C W *et al* 1999 Migration and coalescence of Xe nanoprecipitates in Al induced by electron irradiation at 300 K *Applied Physics Letters* **74**(18) 2611–13; Maiti A *et al* 1999 Damage nucleation and vacancy-induced structural transformation in Si grain boundaries *Applied Physics Letters* **75**(16) 2380–2. American Physical Society: Yan Y and Kirk M A 1998 *Physical Review* B **57** 6152; Birtcher R C and Donelley S E 1996 *Physical Review Letters* **77** 4374; Gibson J M and Treacy M M J 1977 *Physical Review Letters* **78** 1074. Blackwell Science: Jenkins M L *et al* 1973 *Journal of Microscopy* **98** 155, figures A, B, C, D, E, F; Titchmarsh J M and Dumbill S 1997 *Journal of Microscopy* **188** 224. Elsevier Science: Howe L M 1980 *Nuclear Instruments and Methods in Physics Research* **170** 419; Kirk M A *et al* 2000 *Journal of Nuclear Materials* **276** 50; Kiritani M 1994 *Journal of Nuclear Materials* **216** 220; Fukushima H 1994 *Journal of Nuclear Materials* **212–215** 154; Jenkins M L 1994 *Journal of Nuclear Materials* **216** 124; Sagaradze V V *et al* 1999 *Journal of Nuclear Materials* **274** 287; English C A and Jenkins M L 1981 *Journal of Nuclear Materials* **96** 341; Howe L M and Rainville M H 1977 *Journal of Nuclear Materials* **68** 215; King W E *et al* 1983 *Journal of Nuclear Materials* **117** 12; Kiritani M *et al* 1993 *Journal of Nuclear Materials* **205** 460; Williams T M *et al* 1982 *Journal of Nuclear Materials* **107** 222; Shepherd C M 1990 *Journal of Nuclear Materials* **175** 170; Titchmarsh J M and Dumbill S 1996 *Journal of Nuclear Materials* **227** 203. Gordon and Breach Publishers: Ruhle M and Wilkens M 1975 *Radiation Effects and Defects in Solids, Section B: Crystal Lattice Defects and Amorphous Materials* **6** 129. Institute of Physics Publishing: Eyre B L *et al* 1977 *Journal of Physics F: Metal Physics* **7** 1359. Oxford University Press: Jenkins M L *et al* 1999 *Journal of Electron Microscopy* **48** 323. The Royal Society: Saldin D K *et al* 1979 *Phil. Trans. Roy. Soc. Lond.* A **292** 524, figure 13; Bullough T J *et al* 1991 *Proc. Roy. Soc. Lond.* A **435** 85; Black T J *et al* 1987 *Proc. Roy. Soc. Lond.* A **409** 177, figure 1.

Preface

Our motivation for writing this book was two-fold. First, no existing book provides a complete and detailed description of the TEM techniques which have been developed for characterization of small point-defect clusters and other fine-scale radiation damage microstructures. Several good books cover the general theory and practice of transmission electron microscopy (e.g. Hirsch *et al* 1967, Williams and Carter 1996), but none describes fully the range of techniques considered here. These books do of course cover many topics, especially diffraction theory, in more detail than is possible in this book. Second, we fear that important lessons from the past, especially about some of the inherent difficulties and limitations of some of the techniques, are in danger of being forgotten.

We hope for these reasons that this book will be of practical use to microscopists working in the field of radiation damage. Several of the techniques described should also be useful more generally, in applications where it is necessary to characterize fine-scale, complex microstructures. We will show, for example, some studies of ion-implantation damage in semiconductors and flux pinning defects in superconductors.

We are indebted to a large number of our friends and colleagues with whom we have worked over the years. We would mention particularly Colin English, Ian Robertson and Hiroshi Fukushima. We are grateful to Ian, Hiroshi, Brian Eyre, Michio Kiritani, Murray Gibson, Bob Birtcher, Charles Allen, Stephen Pennycook, Paul Okamoto and Robin Schäublin who all contributed original micrographs.

Most of our own research and publications in this field comes through our association with many talented and hard-working graduate students and postdoctorates over the past 20 years. For this we wish to thank Mike Bench, Tom Black, Tim Chandler, Ty Daulton, Marcus Frischherz, Graham Hardy, Simon King, Phil Knight, Bill Lee, Alison Nicol, Peter Othen, Jonathan Perks, Ian Robertson, Martin Robinson, Barry Shepherd, Brad Storey, Czes Pienkowski, John Vetrano, Bob Wheeler, and Yong Yan.

Some of the material, particularly in chapters 2–5, has been published in a similar form in a review by one of the authors (Jenkins 1994). We are grateful to Elsevier Press for permission to reproduce this material here. Many other figures have been taken from the literature, and we are grateful to the various publishers for permission to use these. The sources of the figures are stated in figure captions. We are also grateful to our home institutions, the Department of Materials at the University of Oxford and the Materials Science Division at Argonne National Laboratory, for support in this and past endeavours during our frequent mutual visits. Funding support for MAK has come primarily from the US Department of Energy, Office of Science.

Finally, we thank our families, especially our long-suffering wives Felicity and Stephanie, for their encouragement and support.

Mike Jenkins
Mark Kirk
Oxford, May 2000

Chapter 1

The role of transmission electron microscopy in characterizing radiation damage

1.1 Introduction

Transmission electron microscopy (TEM) is probably the most important and widely used method of characterization in materials science. It has been particularly important in the field of radiation effects, where it has made many essential contributions to our understanding of how microstructures develop under irradiation. This book has two main aims: (1) to describe in detail the methods of transmission electron microscopy used to investigate radiation-induced microstructures, especially small point-defect clusters; and (2) to describe selected examples from the research literature which illustrate the contribution of TEM to our present understanding of radiation-damage mechanisms.

1.2 What is radiation damage?

There are a number of operational environments where the performance of materials is likely to be affected significantly by fast-particle irradiation. These include amongst others: fission reactors, proposed fusion reactors, nuclear waste storage containers, particle accelerators, and spacecraft. An understanding of the basic processes involved when materials are exposed to energetic particle irradiation is clearly an important aspect in the choice of materials for use in such environments. There are also cases where irradiation is used as a processing tool, such as ion-beam modification of surfaces and ion-implantation of semiconductors, where again an understanding of fundamental damage mechanisms is of importance. Traditionally, however, the subject has been driven by the needs of the nuclear power industry. As early as 1942, E P Wigner realized that fast-neutron irradiation of crystalline materials would cause displacements of lattice atoms, and so was likely to affect the physical properties of the uranium fuel and the graphite moderator of the Hanford reactors then under development. Reactor structural components would be likewise affected. These predictions were amply borne out in practice when nuclear fission power reactors were developed, so that the effects of irradiation on structural materials became a subject of great technological importance. A large number of different radiation effects have now been identified. Most of these effects, for example radiation embrittlement and void swelling, are deleterious to the performance of the material: hence the term *radiation damage*.

1.3 Why is electron microscopy useful for studying radiation damage?

The macroscopic effects of irradiation are consequences of events occurring on an atomic scale, and if these effects are to be understood it is necessary to understand how the microstructures of irradiated materials evolve. A number of strategies have been deployed to investigate fundamental damage mechanisms. Since reactor structural materials such as steels are very complex, simpler model materials such as pure elements or binary alloys are often studied. In a reactor damage accumulates relatively slowly, and materials become activated and therefore difficult to handle. Neutron irradiation is therefore often simulated by electron or ion irradiation, where much higher dose rates are possible, and the influence of materials and irradiation parameters can be explored more systematically. A wide variety of irradiated materials including many metals and alloys, ceramics, semiconductors, and superconductors have now been studied using a large range of techniques. It has become clear in such studies that radiation-damage mechanisms are generally very complicated. Irradiation may induce both structural and compositional changes. The resulting radiation-damage microstructures are among the most complex known. Even in pure materials such as copper, various point-defect clusters of different natures, sizes and types may be present. Depending upon the irradiation conditions and material, such clusters may include dislocation loops of vacancy or interstitial nature, stacking-fault tetrahedra, bubbles and voids, all in a size range of the order of nanometres. Line dislocation densities may be high and the dislocation geometries complicated. In alloys, precipitation of second phases may be induced, enhanced, suppressed or impeded. Segregation of alloying elements to or away from grain boundaries or dislocations may occur.

In the analysis of such complex microstructures and their interpretation in terms of fundamental radiation-damage mechanisms, TEM has played a central role. It has the priceless advantage of being very direct. Complementary techniques such as x-ray and neutron diffraction, and many others, can give very useful information on cluster populations, and are often essential for a full characterization of microstructures. Some of the new microscopy techniques, such as the position-sensitive atom-probe, have a useful niche role in radiation-damage studies, but are not generally applicable to a wide range of characterization problems. However, TEM is capable of routine direct imaging at magnifications of the order of 10^5 or higher at a resolution better than 1 nm. This allows you to 'see' individual defect clusters, and 'seeing is believing'. Combined with *in situ* analytical techniques, TEM can also give information on local chemical changes on a very fine scale. Properly applied, TEM can give insight into mechanisms of damage evolution which no other technique can deliver.

A few examples from the early research will serve to demonstrate how TEM immediately became indispensable in the radiation-damage field. In 1959, Silcox and Hirsch made the first observation of defects produced by neutron irradiation in any material. From their work in copper it was immediately obvious that electron microscopy would play a central role in elucidating the defect structures resulting from irradiation of metals. This began perhaps the most productive decade of research in both the structure of irradiation defects and the development of TEM techniques to analyse this structure. The latter was advanced significantly by the groups led by L M Brown at Cambridge University, by B L Eyre and R Bullough at Harwell, and by M Wilkens and M Rühle at the Max Plank Institüt für Metallforschung in Stuttgart. Classic papers by Ashby and Brown (1963), Rühle *et al* (1965), Maher *et al* (1971) and Eyre *et al* (1977a, b) were instrumental in providing methods for the TEM analysis of small (<10 nm) defects produced by irradiation. Merkle *et al* (1963, 1966) took advantage of some of these techniques to first show the differences in defect sizes, morphologies and densities due to different irradiating particles and energies. At about the same time period, Cawthorne and Fulton (1966) showed by TEM methods the formation of voids in neutron-irradiated steel, a discovery to have significant impact on the design and materials used in subsequent nuclear reactors. Another significant discovery of a fundamental irradiation effect using TEM techniques was that of Okamoto and Wiedersich (1974) who showed the first experimental evidence of, and proposed, the first physical explanation for, radiation-induced segregation (RIS). This idea of RIS was to prove essential in understanding the structural and compositional responses of complex steels and alloys to neutron irradiation.

A particularly valuable feature of TEM is the ability, in some instruments, to carry out *in situ* irradiations with electrons or ions of sufficient energy to cause atomic displacements. Such instruments allow the accumulation of damage to be observed directly as it occurs, with good control over the irradiation temperature and dose. In metals, the threshold electron energy for causing displacements is of the order of a few hundred keV, depending on atomic number and the orientation of the specimen with respect to the electron beam. In copper, for example, this threshold is about 400 keV at the $\langle 111 \rangle$ orientation (King *et al* 1983). If the electron microscope operates at a voltage higher than this, then the beam electrons will cause irradiation damage in the area under observation. This possibility was one of the main rationales for the development of so-called high-voltage electron microscopes, which typically operate at 1 MeV or more. A few instruments (such as the ones at Argonne National Laboratory which we have used extensively, see chapter 9) also have the capacity for *in situ* ion irradiations.

1.4 The limitations of TEM

In common with all other experimental techniques, TEM has its limitations and these limitations must be recognized and understood if reliable information is to be obtained. The foils used in TEM are thin, usually less than, and sometimes much less than, 1 μm. Thin-foil artefacts of various types can occur. The situation is least serious for specimens irradiated in the bulk, but even in this case artefacts may be introduced during thin-foil preparation. Some examples will be described in section 2.5.

Other limitations arise from the imaging process in the TEM itself. The resolution is limited, so very small defect clusters may not be seen. What is seen may depend sensitively on the experimental conditions which are set up. How to deal with this is another of the main themes of this book.

In practice, it is often advantageous to combine electron microscopy with other characterization techniques. In this way it is possible to build up a more complete picture of the microstructure than any one technique can give. There are many such techniques, and they each have their advantages and disadvantages. For example, the technique of small angle neutron scattering is sensitive to defects below the resolution limit of TEM, and since it samples large volumes it can give good information on defect-cluster size distributions and number densities. However, it is not an imaging technique and so gives little information on local defect morphologies. Important though these other techniques are, they are not the topic of this book and so will not be discussed further.

1.5 The scope of the book

A range of TEM methods has been developed to address the particular problems inherent in characterizing radiation-induced microstructures, especially small point-defect clusters. The main aim of this book will be to describe these techniques in detail. The emphasis will be on the experimental practice of the techniques and how to extract the information which they are capable of delivering. However, we shall present many examples of studies in various materials (including metals, ceramics, semiconductors, and superconductors) and we hope that these will serve to illustrate the contribution of TEM to the present understanding of radiation-damage mechanisms.

The main questions we shall address are these: What needs to be done experimentally to achieve a reliable characterization of radiation-damage microstructures? What can we infer reliably from the images? What pitfalls must we avoid? We shall cover both conventional diffraction-contrast and high-resolution imaging techniques, and, at a less detailed level, techniques of analytical electron microscopy.

1.5.1 Organization of the book

Chapter 2 gives a brief description of the various methods of imaging small point-defect clusters in the TEM. The emphasis is on how to set up appropriate imaging conditions for different purposes, and how to interpret the images. We also briefly discuss data acquisition, and image artefacts.

Chapters 3–5 give detailed descriptions of the available techniques for characterizing small dislocation loops. The determination of loop morphologies—finding Burgers vectors and habit-planes—is described in chapter 3. Chapter 4 deals with both direct and indirect methods for determining the nature, vacancy or interstitial, of loops. Chapter 5 considers quantitative evaluations of loop number densities and sizes.

In chapter 6, characterization of voids and bubbles is discussed. Chapter 7 describes methods for direct characterization of displacement cascades. Chapter 8 considers the application of high-resolution techniques to radiation-damage problems. Chapter 9 gathers together some examples of *in situ* techniques not discussed earlier. Chapter 10 gives a brief description of analytical methods, and looks at some possible future trends. Finally, in chapter 11 we briefly discuss the investigation of radiation damage in amorphous materials.

All of the techniques are described with selected examples to help illustrate the contribution TEM has made to our present understanding of radiation-damage mechanisms.

Chapter 2

An introduction to the available contrast mechanisms and experimental techniques

We shall assume in this book a working knowledge of basic TEM at a level covered for example by Williams and Carter's *Transmission Electron Microscopy* (1996). The contrast of small clusters is very dependent on the contrast mechanism employed. The image quality and the threshold visibility of defects are also strongly dependent on proper focusing and correction of objective astigmatism. This chapter summarizes those contrast mechanisms routinely used in TEM of point-defect clusters and microstructures resulting from radiation damage, and describes some of the experimental procedures used. More complete descriptions of experimental techniques are found later in the book.

2.1 Diffraction (or strain) contrast

In diffraction contrast, defects are imaged by virtue of their elastic strain fields. The crystal is set at some well-defined diffracting condition, and an objective aperture is used to form an image using one of the diffracted beams. The defect strain field causes local changes in diffracting conditions: for example, the diffracting planes may be locally bent, causing changes in the amplitude of the diffracted beam used to form the image. This is the most common mechanism used in radiation-damage studies. The image characteristics depend sensitively on the diffraction conditions chosen. Several different types of condition are used and these are described later. In each case it is very important to set the diffraction condition carefully (and of course to describe the condition clearly in any published micrographs). A full dynamical theory of diffraction contrast is given in the book *Electron Microscopy of Thin Crystals* by Hirsch *et al* (1977), and in many other books, and will not be described here. We will, however, occasionally make references to key results of the dynamical theory.

In this book we shall mostly be concerned with the imaging of small point-defect clusters. 'Small' in this context is in relation to the extinction distance ξ_g, which plays a critical role in the description of electron diffraction phenomena. The value of ξ_g depends on the material, the excited reflection $\boldsymbol{g}$ (see later), and the microscope operating voltage, but is typically several tens of nanometres.

2.1.1 Two-beam dynamical conditions

Under two-beam dynamical conditions, sometimes called 'strong' two-beam conditions, the contrast is sensitive to weak lattice strains, and so these conditions are particularly well-suited to the investigation of the symmetry and sign of the long-range elastic strain fields of small clusters such as dislocation loops.

Dynamical two-beam conditions imply that the foil is tilted so that one set of diffracting planes (hkl) is at, or very close to, the Bragg condition. The Ewald sphere construction for this condition is shown schematically in figure 2.1(a). The reciprocal lattice vector $\boldsymbol{g}$ in the construction corresponds to the *diffraction vector* joining the (000) to (hkl) spot in the diffraction pattern. In the figure, the curvature of the sphere has been increased for clarity; it is actually rather flat (with radius $k = \lambda^{-1}$ where λ is the electron wavelength, so that $k \gg g$). The diffraction condition is often specified by the 'deviation parameter', s_g, which is the distance of the reciprocal lattice point $\boldsymbol{g}$ from the Ewald sphere. The deviation parameter s_g is therefore zero or very small under dynamical two-beam conditions.

The beam $\boldsymbol{g}$ will then appear in the diffraction pattern typically with intensity similar to that of the forward-scattered (or 'incident') beam: these are the 'two beams', see figure 2.1(a). Other beams are always present, but with much smaller intensities. It is impossible to eliminate the weakly-excited systematic reflections $(\ldots -\boldsymbol{g}, 2\boldsymbol{g}, 3\boldsymbol{g}\ldots)$ because of the small curvature of the Ewald sphere and the reciprocal lattice spikes normal to the thin foil. This problem is more important the higher the voltage of the microscope used, since k increases and the curvature of the Ewald sphere decreases. However, it is possible to avoid exciting strongly *non*-systematic reflections by appropriate tilting of the specimen, and care should be taken to ensure this.

The image is formed by placing an objective aperture, situated in the back-focal plane of the objective lens, around either the forward-scattered beam or the diffracted beam to form a bright-field or a dark-field image, respectively. Dark-field images are less affected than bright-field images by the presence of other diffracted beams. For this reason they generally have higher contrast and so are often preferred. In order to obtain such images, the diffracted beam is first tilted onto the optical axis using the beam-tilt coils, so as to avoid loss of resolution due to spherical aberration. This is called a 'centred' dark-field image. The two-beam condition is usually set up by observation of the Kikuchi pattern in the diffraction mode. A commonly-used experimental procedure for attaining a dark-field two-beam condition is first to tilt the foil in bright-field diffraction mode to excite a reflection $\boldsymbol{g}$ (i.e. $s_g = 0$). The dark-field diffraction mode is then selected, and the reflection $-\boldsymbol{g}$ is tilted to the optical axis with the beam-tilt coils. When $\boldsymbol{g}$ is on the axis, it will be fully excited (i.e. $s_{-g} = 0$). This method has the advantage

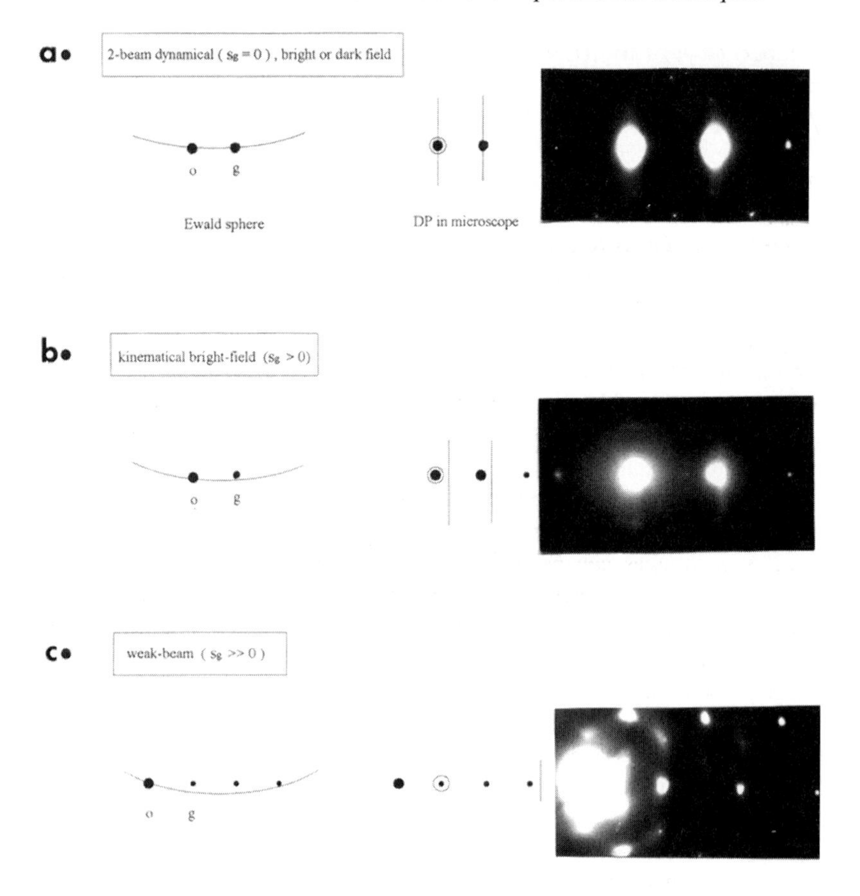

Figure 2.1. Three of the different types of diffraction condition used in diffraction-contrast imaging: (*a*) two-beam dynamical, (*b*) two-beam kinematical, (*c*) weak-beam. In each case the Ewald sphere is sketched on the left-hand side, a schematic diffraction pattern (DP) showing the position of the relevant Kikuchi lines is shown at the centre, and an actual diffraction pattern is shown on the right-hand side. The curvature of the Ewald sphere is exaggerated for clarity. The open circle represents the objective aperture.

that two-beam conditions pertain in both bright-field and dark-field modes, so that two-beam bright-field and dark-field images can be switched immediately without foil tilting. However, it has to be remembered that the operating reflection in the dark-field mode is then $-\boldsymbol{g}$ not $\boldsymbol{g}$, which is important in the determination of the defect natures (see chapter 4). In cases where the foil is bent locally or when the dislocation density is high, it may not be possible to set diffraction conditions using the Kikuchi pattern. It is still sometimes possible to obtain useful images simply by maximizing the intensity of the dark-field image by tilting the sample whilst viewing in dark-field conditions.

Under strong two-beam conditions, there is a strong dynamical interaction between the two beams, and the resulting image characteristics may be quite complicated. Image peak widths are generally rather broad, about $\frac{1}{3}\xi_g$ for a line dislocation. Loops smaller than ξ_g will not be resolved directly, in the sense that their images will not be 'loop-like' but will be more complex. In many cases, for example in studies of line dislocations and their interactions, this is a disadvantage. However, the contrast features shown by small defect clusters under strong two-beam conditions can be exploited usefully. If the foil thickness is greater than about $2\xi_g$, typically about 60–80 nm depending on the reflection and operating voltage, small dislocation loops located close to the foil surfaces show so-called black–white contrast, consisting of pairs of black and white lobes. The symmetry of such images can be used to deduce the Burgers vectors and habit-planes of the loops, and also their nature provided that their depth in the foil can be determined.

An example of defect analysis using black–white contrast is shown in figure 2.2. The method is discussed in detail in chapter 3, with further discussion on how such images can be used to determine the loop nature and to make size measurements in chapters 4 and 5. We emphasize here, however, that in studies of this sort it is *essential* to carry out systematic experiments under well-defined diffracting conditions, using several diffraction vectors in turn. Any one micrograph will generally be insufficient to identify the loop types present, or even to distinguish between loops and other features such as small precipitates or stacking-fault tetrahedra (SFT). However, the *changes* in contrast under different diffraction conditions *are* characteristic of different cluster types. We will refer to the type of experiment illustrated in figure 2.2 as a 'contrast experiment'. In the experiment shown in figure 2.2, the images were obtained using the four reflections available at the [011] specimen orientation. At the exact pole orientation, the crystallographic planes giving rise to these reflections are all edge-on. The desired diffraction conditions for each micrograph were obtained by controlled tilting, such as to excite the reflection of interest and suppress the other reflections. Since Bragg angles in electron diffraction are small (of the order of a degree), this could easily be achieved by tilting a few degrees away from the pole. In more complete contrast experiments it is sometimes necessary to tilt by much larger angles in order to access particular reflections.

2.1.2 Bright-field kinematical conditions

Bright-field kinematical conditions are used when it is desired to avoid the dynamical contrast effects seen in strong two-beam images. Typical bright-field kinematical conditions are shown in figure 2.1(*b*). A two-beam condition is set up with a small positive deviation parameter, i.e. $s_g > 0$. Often the magnitude of the deviation parameter is not specified explicitly. The microscopist simply tilts the foil just sufficiently away from the Bragg condition that the image loses most of its dynamical features. It can be shown that the effective extinction distance is

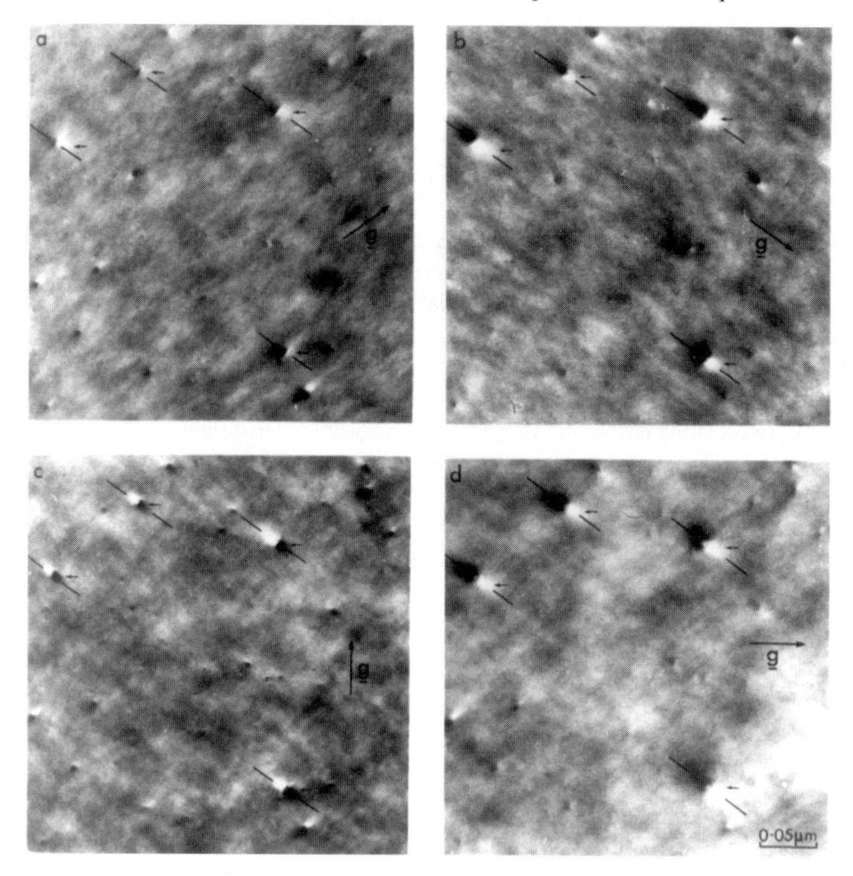

Figure 2.2. Black–white contrast of dislocation loops under dynamical two-beam imaging conditions. This figure shows the same area of a copper specimen, containing small dislocation loops formed by irradiation with 30 keV self-ions. The area is imaged in turn under two-beam dynamical conditions using four different reflections. The $\boldsymbol{g}$-vectors are shown as larger black arrows, and are: (*a*) $\boldsymbol{g} = 1\bar{1}1$, (*b*) $\boldsymbol{g} = \bar{1}\bar{1}1$, (*c*) $\boldsymbol{g} = 200$, and (*d*) $\boldsymbol{g} = 0\bar{2}2$. In each micrograph, only the reflection indicated is strongly excited. This is achieved in practice by controlled tilting of the foil, in this case in the neighbourhood of the [011] pole. By following the contrast changes shown by the loops from one micrograph to another the loop Burgers vectors can be determined. This is discussed in detail in chapter 3. The arrowed defects are in fact edge-on Frank loops with $\boldsymbol{b} = \frac{1}{3}[11\bar{1}]$. The lines indicate the projection $\boldsymbol{b}_{\rm p}$ of the loop Burgers vector on the image plane. Micrographs courtesy of A Y Stathopoulos (1977, see also Stathopoulos 1981).

reduced to

$$\xi_g^{\rm eff} = \xi_g/\sqrt{(1+w^2)}$$

where $w = s_g \xi_g$ is a dimensionless deviation parameter, and so images also become narrower. The contrast is then expected to be more confined to the physical size of the cluster. An example is shown in figure 2.3, taken from Sigle *et al* (1988). Here a dislocation in silver is visible as a black line, and small stacking-fault tetrahedra, which have formed in its vicinity under 1 MeV electron irradiation, are visible as black triangles. The simplified contrast under such conditions offers considerable advantages. For this reason, bright-field kinematical conditions have often been used to image small loops. The contrast changes of the same field of small Frank loops as the dimensionless deviation parameter w is increased from zero to one is shown in figure 2.4. As w increases the images lose their dynamical character becoming simple black dots. However, quantitative measurements, such as measuring sizes, should be made from such images only with caution (see chapter 5).

2.1.3 Down-zone conditions

Another type of image results when the foil is tilted to a zone-axis orientation, so that several diffracted beams are more or less equally excited. Images obtained in this way were sometimes termed 'kinematical' in the early literature, but are now more often termed 'down-axis images' or 'bright-field down-zone images'. This condition offers potential advantages in determinations of the loop number density, since a high proportion of all loops present are expected to be imaged in a single micrograph. An example of down-zone imaging of small clusters in ion-irradiated copper is shown in figure 2.5(*a*). In practice it has been found that weak-beam microscopy (which is discussed later) is superior for identifying very small clusters, see figure 2.5(*b*), and chapters 3 and 5. Also, there is the same need for caution regarding quantitative size measurements, see chapter 5.

Down-zone imaging has been shown to be very useful for imaging amorphous zones in semiconductors and superconductors, where the contrast mechanism is structure factor contrast, see figure 7.9 and Bench *et al* (1992).

2.1.4 Weak-beam dark-field conditions

As the name implies, weak-beam images are dark-field images obtained using a weakly-excited beam. The foil is tilted so that in regions remote from dislocations or other defects, the diffracting planes are well away from the Bragg condition (figures 2.1(*c*) and 2.6). The magnitude of the deviation parameter $|s_g|$ is therefore large. Although figure 2.1(*c*) shows s_g as positive, in practice s_g can be either positive or negative. The contrast under weak-beam conditions arises from regions of large lattice strain close to the cores of defects. Since the foil is tilted well away from the Bragg condition the average image intensity is very low. However, in regions close to defects, the local strain-field may bend the reflecting planes back towards the Bragg condition. This leads to a substantial enhancement of the defect peak intensity relative to the surrounding background, resulting in a

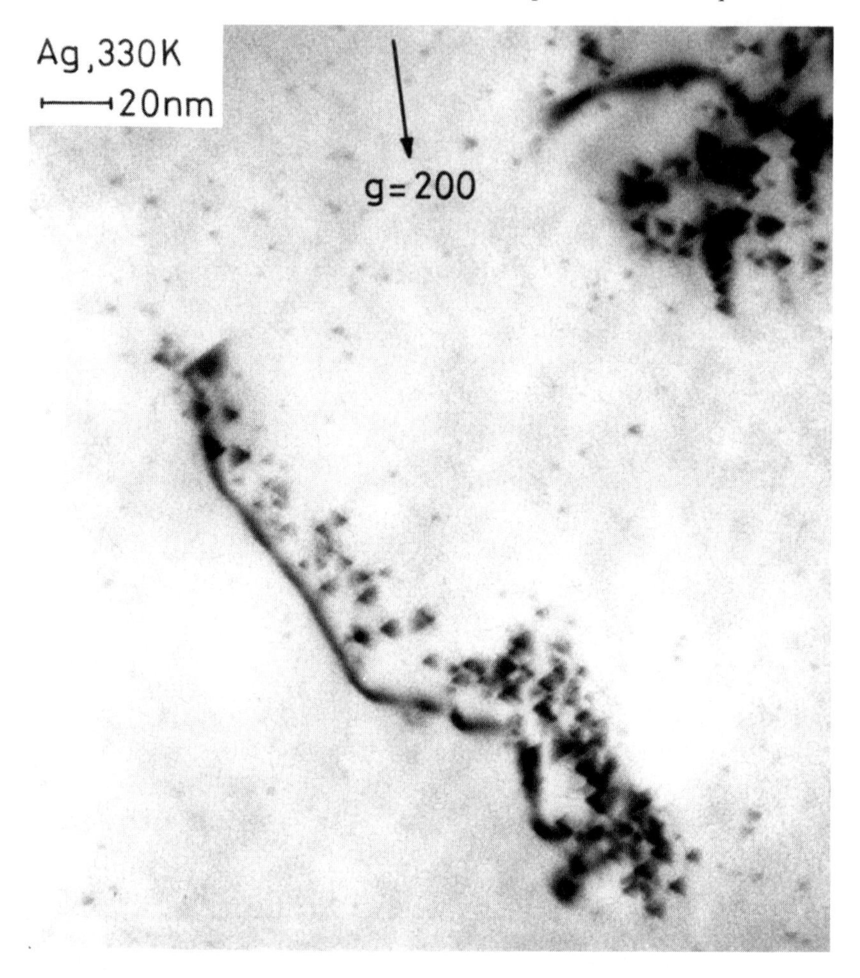

Figure 2.3. Stacking-fault tetrahedra (SFT) near line dislocations in silver imaged under kinematical conditions. In this projection the SFT appear as small black triangles. They were formed under 1 MeV electron irradiation at 330 K. Such electrons have sufficient energy to displace lattice atoms in bulk crystal, creating Frenkel pairs. Vacancies then agglomerate to SFT in regions of compressional strain near dislocations. From Sigle *et al* (1988).

local image peak of high relative (but low absolute) intensity.

It is generally considered that for quantitative measurements from an interpretable image it is necessary to set a value of the deviation parameter

$$|s_g| \geq 2 \times 10^{-1}\ \mathrm{nm}^{-1}.$$

For imaging of line dislocations, this condition results in a narrow (<2 nm),

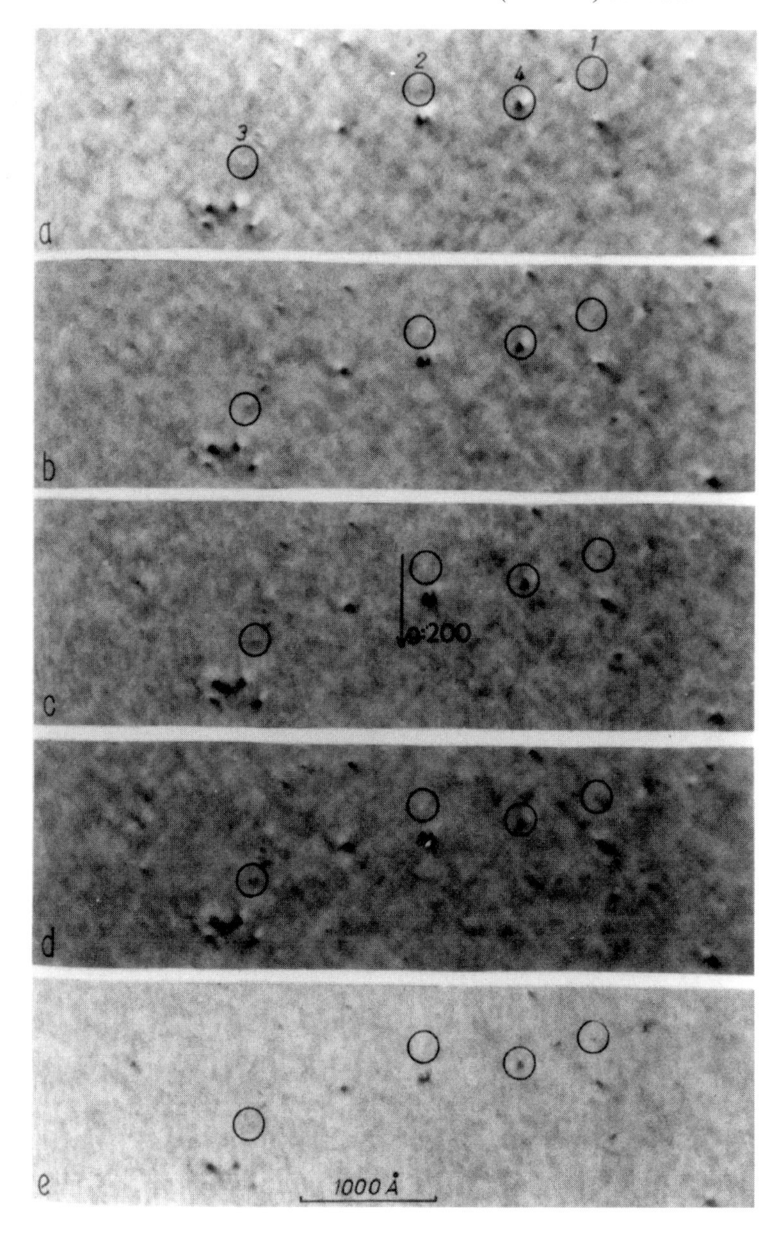

Figure 2.4. Images of small point-defect clusters, mostly Frank dislocation loops, as a function of the dimensionless deviation parameter $w = s_g \xi_g$. The micrographs show the same area of a copper foil irradiated with 30 keV Cu^+ ions, imaged in a bright-field with $\boldsymbol{g} = 200$, and with: (*a*) $w = 0$ (the exact Bragg condition); (*b*) $w = 0.2$; (*c*) $w = 0.4$; (*d*) $w = 0.6$; (*e*) $w = 1$. As the images become more kinematical, the black–white contrast changes to black-dot contrast. Micrographs courtesy of Dr A Y Stathopoulos (1977).

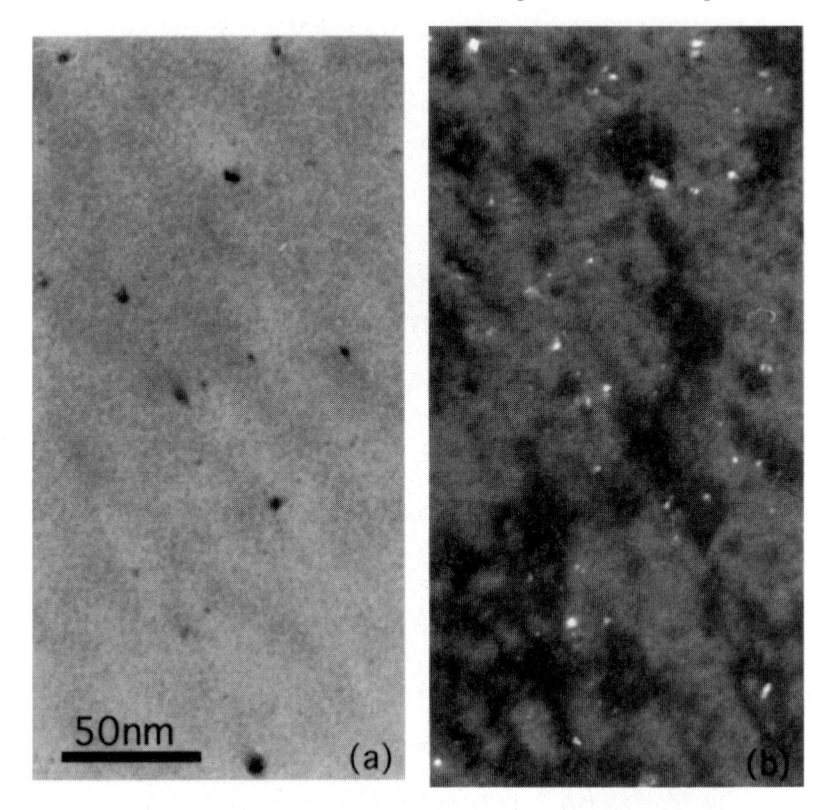

Figure 2.5. The same area of a copper foil irradiated with 600 keV Cu^+ ions, imaged: (*a*) in a bright-field 'down-zone condition', with the foil accurately tilted to the [110] pole; (*b*) in a weak-beam condition, with $\boldsymbol{g} = 002$ and $s_g = 2 \times 10^{-1}$ nm^{-1}. It is clear that defects show better visibility in the weak-beam condition.

high-contrast image peak located close to the dislocation core (Cockayne 1973). The same value of $|s_g|$ has also been found to be suitable for imaging small point-defect clusters (Jenkins *et al* 1973, 1999a). The diffraction condition needed to achieve this value of $|s_g|$ can be set by reference to Kikuchi lines in the diffraction pattern. Suppose the foil is tilted such that the Ewald sphere cuts the line of systematic reflections at the point $n\boldsymbol{g}$ (figure 2.1(*c*)). The value of n is immediately apparent from the Kikuchi pattern. If n is integral then the bright excess Kikuchi line corresponding to the reflection $n\boldsymbol{g}$ will pass directly through the diffraction spot $n\boldsymbol{g}$. However, n need not be integral. Indeed n *should not* be integral, see later. In this case the Kikuchi line corresponding to the reflection which is closest to being excited will pass close to but not through the diffraction spot, and from its position n can be estimated. This is the case in figure 2.6 which shows a weak-beam diffraction pattern where $n = 5.5$.

Figure 2.6. Experimental diffraction pattern showing a foil oriented at a $(\boldsymbol{g}, 5.5\boldsymbol{g})$ weak-beam diffraction condition. The specimen is silicon, and the pattern was obtained at a microscope operating voltage of 100 kV with $\boldsymbol{g} = 220$. This gives a value for the deviation parameter $s_g = 2.2 \times 10^{-1}$ nm^{-1}.

The required value of n for a given s_g, or *vice versa*, can be found with the aid of figure 2.7. This schematic diagram in reciprocal space of the Ewald sphere, and vectors pertinent to diffraction from a specimen tilted off the Bragg condition at $\boldsymbol{g}$, is not drawn to scale. For clarity the Ewald sphere is drawn with a radius (k) about 50 times smaller than it should be with respect to $|\boldsymbol{g}|$. Using the theorem of

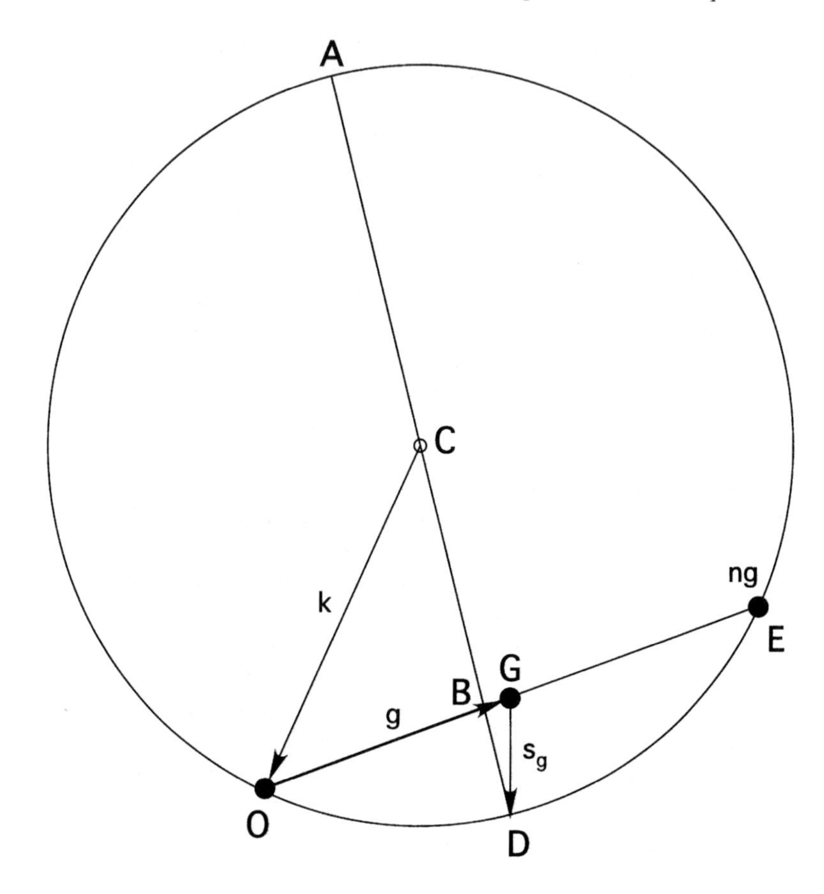

Figure 2.7. Calculation of the deviation parameter s_g from the diffraction pattern. The image is obtained in the reflection $\boldsymbol{g}$, and the Ewald sphere cuts the line of systematic reflections at the point $n\boldsymbol{g}$, where n need not be integral. The deviation parameter, s_g = GD, is in a direction bisecting the incident beam CO and the diffracted beam CD. Since, however, the Ewald sphere is drawn with a radius (k) about 50 times smaller than it should be with respect to $|\boldsymbol{g}|$, GD is in fact closely parallel to the chord BD. To a good approximation therefore BD $= s_g$. The theorem of intersecting chords states that OB × BE = AB × BD. Since OB ≈ OG $= |\boldsymbol{g}| = g$, BE $\approx (n-1)g$, and AB $\approx 2k - s_g$, $(n-1)g^2 = (2k - s_g)s_g$. The equation given in the text follows by neglecting the small term in s_g^2.

intersecting chords the following relation between n and s_g can then be derived:

$$s_g = (n-1)g^2/2k$$

where $g = 1/d_{hkl}$ is the magnitude of the reciprocal lattice vector $\boldsymbol{g}$

corresponding to the planes (hkl) with interplanar spacing d_{hkl}; $k = 1/\lambda$ is the radius of the Ewald sphere, where λ is the electron wavelength; and the 'reflection' $n\boldsymbol{g}$ is satisfied in the sense explained earlier. This condition is often denoted $(\boldsymbol{g}, n\boldsymbol{g})$.

Two common misconceptions about weak-beam imaging should be noted:

- First, it is *not* necessary to excite some higher-order reflection such as $3\boldsymbol{g}$, i.e. setting $n = 3$. Indeed it is important not to do so. Early work (e.g. Cockayne *et al* 1969) did recommend $(\boldsymbol{g}, 3\boldsymbol{g})$ or $(\boldsymbol{g}, -\boldsymbol{g})$ conditions, but this was done only to achieve a sufficiently large value of $|s_g|$ and before it was realized that the presence of strongly-excited systematic reflections could cause image artefacts (Cockayne 1973). In practice one should tilt the foil somewhat away from the exact Bragg condition for the higher-order reflection, so that n is, say, 3.1 rather than exactly 3.
- Second, $(\boldsymbol{g}, \sim 3\boldsymbol{g})$ conditions are not always appropriate. Some authors invariably use $(\boldsymbol{g}, \sim 3\boldsymbol{g})$ conditions irrespective of material, reflection or operating voltage. This again seems to stem from early studies. For a {220} reflection in copper at a microscope operating voltage of 100 kV, a $(\boldsymbol{g}, 3\boldsymbol{g})$ condition does indeed give $s_g = 2 \times 10^{-1}$ nm^{-1}. For a {111} reflection at 200 kV however, the appropriate condition is $(\boldsymbol{g}, \sim 7\boldsymbol{g})$ and $(\boldsymbol{g}, 3\boldsymbol{g})$ gives only $s_g = 0.6 \times 10^{-1}$ nm^{-1}. Images obtained using a value of $s_g < 2 \times 10^{-1}$ nm^{-1} but such that $w = |s_g \xi_g| \geq 5$ are sometimes referred to as semi-weak beam images. This condition results for a line dislocation in a narrow image peak of high relative intensity, but the image peak position is not well defined. Such conditions should therefore generally be avoided for quantitative work. Conditions with $w < 5$ should always be avoided.

The narrow image peaks obtained in weak-beam microscopy are a consequence of the small value of the effective extinction length ξ_g^{eff}.

For $w = s_g \xi_g \geq 5$,

$$\xi_g^{\text{eff}} = \xi_g / \sqrt{(1 + w^2)} \approx (s_g)^{-1}.$$

With $s_g = 2 \times 10^{-1}$ nm^{-1}, $\xi_g^{\text{eff}} \approx 5$ nm, giving dislocation peak widths of about $\frac{1}{3}\xi_g^{\text{eff}} \approx 2$ nm.

Because of these small dislocation peak widths, weak-beam microscopy has been particularly successful in characterizing dislocation microstructures, including the complex dislocation microstructures frequently produced by irradiation. For example, low-dose neutron irradiation of copper, copper alloys and iron sometimes leads to the production of 'high-damage regions', which consist of dense tangles of dislocations decorated with small loops (e.g. Shepherd *et al* 1987, Jenkins and Hardy 1987, Robertson *et al* 1982). An example of a weak-beam analysis of a high-damage region in a neutron-irradiated copper–germanium alloy is shown in figure 2.8. Complex microstructures are also found

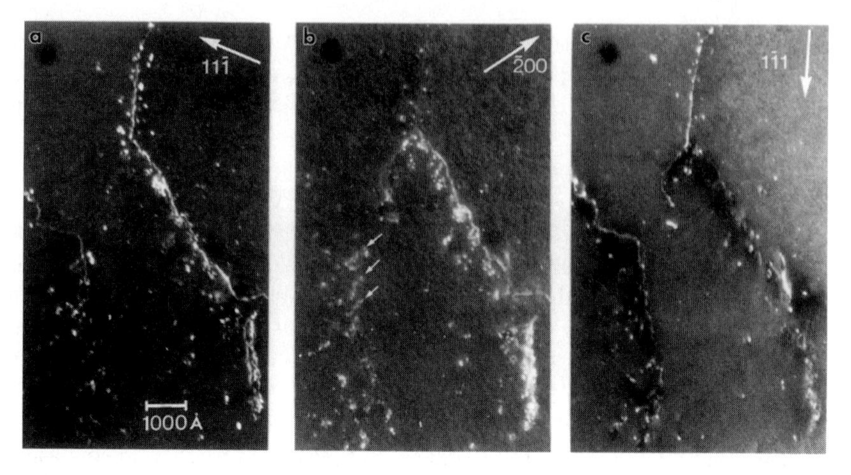

Figure 2.8. Weak-beam images of a 'high-damage region' in a neutron-irradiated copper–germanium alloy, obtained using (*a*) $\boldsymbol{g} = 11\bar{1}$; (*b*) $\boldsymbol{g} = \bar{2}00$; and (*c*) $\boldsymbol{g} = 1\bar{1}1$. Weak-beam microscopy has been used very successfully to characterize complex dislocation microstructures, such as occur in 'high-damage' regions in neutron-irradiated metals under some conditions. These regions consist of tangles of line dislocations decorated with dislocation loops. This figure shows part of a weak-beam analysis of a high-damage region in Cu 0.01% Ge, irradiated with neutrons to a dose of 7×10^{21} m^{-2} ($E > 1$ MeV) at 250 °C. The Burgers vectors and loops can be analysed by standard contrast analysis. Thus, for example, the line dislocation at the top centre is out of contrast in $\boldsymbol{g} = \bar{2}00$, and can be shown to have Burgers vector $\boldsymbol{b} = \frac{1}{2}[01\bar{1}]$ when $\boldsymbol{g} \cdot \boldsymbol{b} = 0$. Note that the loops decorating this dislocation are not out-of-contrast in this reflection, and so do not have the same Burgers vector as the parent dislocation. In fact, these loops have $\boldsymbol{b} = \frac{1}{2}[1\bar{1}0]$. Figure from Jenkins and Hardy (1987).

in studies of the interaction of point-defects with line dislocations under electron irradiation (see e.g. Cherns *et al* 1980, Jenkins *et al* 1987, King *et al* 1993).

The following practical points should be noted:

- It is important to carry out *systematic* contrast experiments, using several different reflections and if necessary varying s_g. Single micrographs can be misleading. Only by a careful comparison of several different micrographs taken under different diffraction conditions can a complete picture of the microstructure be built up. In imaging line dislocations conditions giving $|\boldsymbol{g} \cdot \boldsymbol{b}| > 2$ should be avoided since otherwise spurious image peaks may result.
- It is possible to record good weak-beam micrographs with practically any transmission electron microscope. The technique rarely if ever stretches the electron-optical resolution of the microscope. More important usually is the stability of the specimen stage. Even then, drift problems can often

be overcome by the use of sensitive films and special film development procedures. The exposure time for the micrographs in figure 2.8 was only 8 s. This was achieved by the use of Kodak SO163 film, and by development in concentrated developer (full strength D19) for 10 min. This enabled the exposure time to be decreased by a factor of four compared with a standard electron film and normal development. It is nearly always the case that exposure times in weak-beam microscopy can be kept below 10 s. The general availability of image-intensified, video-rate cameras has made focusing and stigmating of weak-beam images straightforward. An example of an *in situ* weak-beam experiment on point-defect interactions with dislocations in silver carried out on an old HVEM is described in chapter 9, section 9.2.

- Weak-beam images are almost invariably recorded in regions of foil of thickness $t \leq 2\xi_g$. The foil thickness in figure 2.8 was only about 50 nm. In thicker areas the image quality deteriorates quickly due to chromatic aberration. The necessity to use such thin foils is a limitation of the technique, although it should be noted that high-resolution imaging requires the use of still thinner foils. Even this limitation may well be overcome by the advent of energy-filtered imaging in modern field-emission gun microscopes. Preliminary experiments in ion-irradiated copper have shown that weak-beam imaging with zero-loss electrons results in a much improved peak-to-background ratio, allowing defects of size $\leq$1 nm to be identified (see section 3.3). It may be possible to carry out weak-beam imaging in thicker areas by using this technique.
- Kikuchi lines may be difficult to see in conventional selected-area diffraction patterns in relatively thin areas of foil ($t < 60$ nm), making it difficult to set diffraction conditions accurately. We have found that by using a fully-focused beam on the area of interest, and forming a diffraction pattern *without* the use of the selected-area aperture, the visibility of Kikuchi patterns in flat thin foils is much improved.

Weak-beam imaging has been used to analyse very successfully not only line dislocations, but also dislocation loops and other small centres of strain. The superiority of weak-beam imaging over down-zone imaging for imaging small dislocation loops is apparent from figure 2.5. We have shown weak-beam imaging to be capable of imaging 1–2 nm defects throughout a foil of thickness ~60 nm. The detailed procedures we have developed for doing this will be discussed in sections 3.2.2 and 5.2.1.

2.2 Structure-factor contrast

Contrast arises because the damaged regions (voids, bubbles, amorphous or disordered zones) have a different structure factor, and hence a different extinction distance, from the surrounding matrix. For in-focus imaging of cavities or amorphous zones the contrast can be regarded to arise from a different effective foil thickness for columns passing through the defect compared with adjacent columns which pass through perfect crystal[1]. The contrast is therefore at a maximum in the flanks of thickness fringes, where the image intensity changes most rapidly as a function of foil thickness. Imaging of voids and bubbles is considered in more detail in chapter 6. Structure-factor techniques for the direct imaging of displacement cascades are described in chapter 7.

An example of an image dominated by structure-factor contrast is shown in figure 2.9. This micrograph shows images of amorphous zones in silicon produced by heavy-ion bombardment with 60 keV Bi_2 ions. On the thinner side of the thickness fringe the defects appear as dark dots against a white background, whereas on the thicker side of the fringe this contrast is reversed. Similar work in GaAs is reported in chapter 7. In-focus images of voids look much the same (see chapter 6).

2.3 Phase contrast

The specimen introduces local changes in the phase of the electron wave. These local phase variations are converted into amplitude variations via the Scherzer phase shift produced by the microscope over which the operator has control by defocusing. An example is the imaging of voids in out-of-focus conditions (see chapter 6).

An important sub-area of phase contrast imaging is high-resolution structural imaging, which is discussed in detail in the book by Spence (1988). Under some conditions the resulting image may show resolution on a scale close to the atomic level. Generally, however, if the image is to resemble the projected charge density of the specimen, several stringent imaging conditions must be met.

Unlike diffraction contrast images, high-resolution 'structural' images are obtained by allowing several diffracted beams to contribute to the image. Usually, the specimen is tilted to a low-index 'zone-axis' orientation. Then a large objective aperture is used to select a number of beams to form the image[2]. The image is recorded under a carefully selected defocus condition, and great care is taken in correcting astigmatism. This is made easier on modern instruments by the provision of a high-magnification TV system, or even on-line Fourier analysis. Exposure times are kept to a minimum by the use of sensitive films and accelerated film development, as for weak-beam imaging, or intensified video or CCD cameras. The resulting 'structural' image *may*, under certain conditions, resemble the projected charge density of the object.

[1] The meaning of 'column' here is discussed in chapter 3.

[2] In principle no objective aperture is necessary at all but there is no point in including in the image spatial frequencies u beyond the ultimate resolution limit of the microscope, because they simply add noise. Here $u = \theta/\lambda$, where θ is the scattering angle and λ the electron wavelength. The aperture defines a cut-off angle θ_0, corresponding to a spatial frequency $u_0 = \theta_0/\lambda$. See chapter 8.

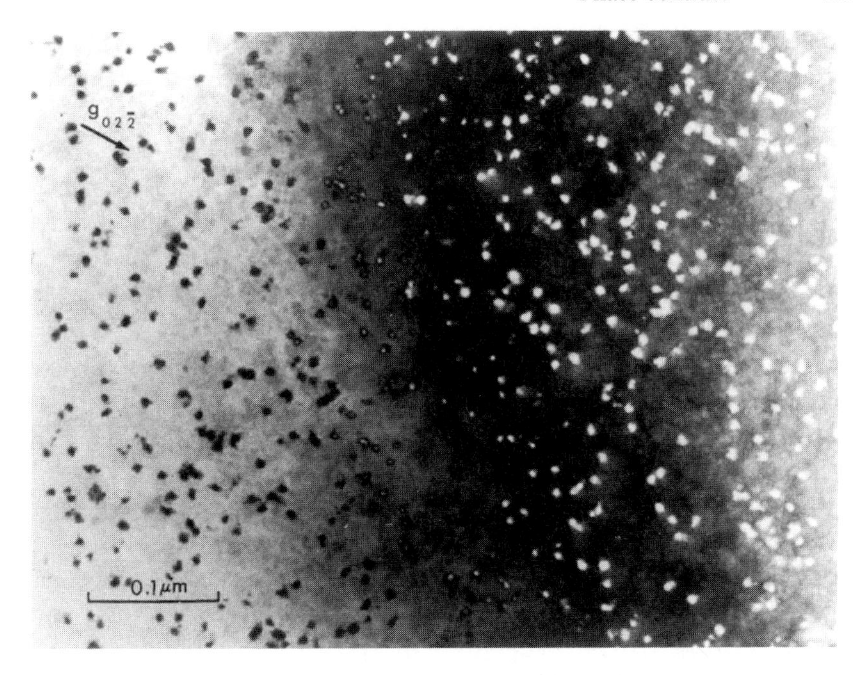

Figure 2.9. Amorphous zones in Si imaged in structure-factor contrast. These were produced by a process of direct-impact amorphization in an irradiation with 60 keV Bi_2 ions. The image was obtained in a dark field under dynamical two-beam conditions, using $g = 02\bar{2}$. The contrast of the amorphous zones arises from an effective local change in foil thickness—see text for details. Note that in the transition region at the centre of the thickness fringe, some zones image as dark dots with a white centre. Figure taken from work by Howe *et al* (1980).

High-resolution imaging has had a limited but important role in investigations of radiation damage. In general, it is not possible to achieve true structural images of small point-defect clusters because they lie buried within the matrix. Even so, it is sometimes possible to obtain useful information. This is discussed in more detail in chapter 8. The potential of the technique is illustrated in figure 2.10. The micrograph in figure 2.10(*a*) shows a high-resolution image grabbed from a video tape of small solid Xe nanocrystalline 'bubbles' in an aluminium matrix under 1 MeV electron irradiation, taken from the work of Allen *et al* (1994, 1999). A special imaging method has been used to 'detune' the contrast from the matrix, so that only the Xe crystallites are visible. The small Xe precipitate at the centre of the field of view is a nearly-perfect cubo-octahedron, a model of which is shown in figure 2.10(*b*) in the appropriate projection. The 14 Xe atom columns of the model are each composed of two to four Xe atoms; the total number of atoms in the model is 38. There is a one-to-one correspondence between the experimental image and the projected structure of the model, showing

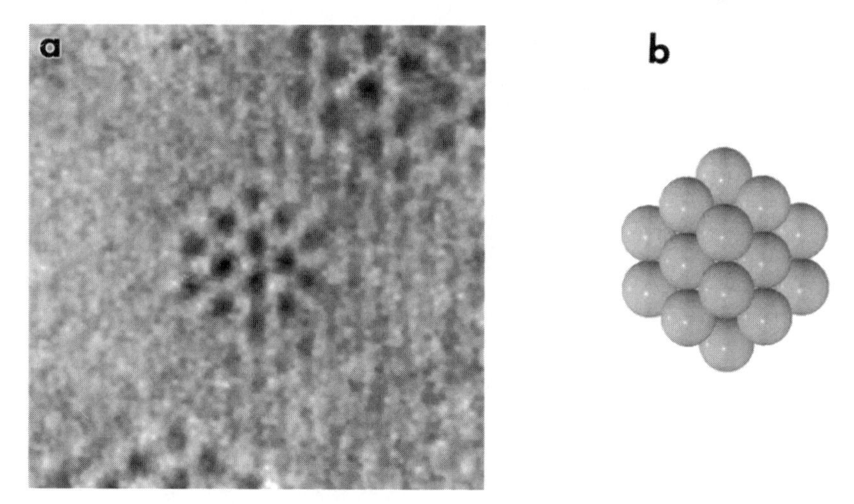

Figure 2.10. High-resolution imaging of a small solid Xe precipitate in aluminium under 1 MeV electron irradiation. The image (*a*) was grabbed from video tape. The imaging technique involved tilting the specimen a few degrees off a ⟨110⟩ zone axis with a non-Scherzer objective defocus, and is discussed in section 8.3. A model of the cubo-octahedral precipitate in a ⟨110⟩ projection is shown in (*b*). The resemblance between (*a*) and (*b*) is striking, and shows that under suitable imaging conditions columns containing as few as two Xe atoms can be imaged successfully. The images were obtained in the JEOL ARM-1000 high-resolution electron microscope in the National Research Institute for Metals in Japan. Figure from Allen *et al* (1999).

that atomic columns consisting of as few as two atoms are well resolved. Further details will be found in section 8.3.5.

An alternative high-resolution imaging technique, high-angle annular dark-field or 'Z-contrast' imaging, does not rely on phase contrast and so gives more easily interpretable images in some cases. This technique is also capable of giving chemical information. It will be discussed briefly in chapter 10.

2.4 A few words on image recording

Conventional TEM images have long been recorded on negative film using, for example, Kodak SO163 for its good compromise between speed and resolution. Digital charge coupled device (CCD) cameras are now becoming increasingly common, particularly for high-resolution electron microscopy (HREM), and for energy-loss imaging using in-column or post-column imaging filters. It is of interest, therefore, to compare recording media in the context of the microscopy of small defects using the techniques discussed in this book.

The main points of comparison between different recording media are their spatial resolution, depth and linearity of grey scale, and field of view. The spatial resolution of a digital camera is defined in large part by the pixel size of the CCD chip, which is typically about 24 μm square. An effective pixel size for film, as recently defined by Eades (1996), is estimated at nearly the same value, 25 μm, which is similar to the pixel size (21 μm) of a negative (film) scanner operating at 1200 dpi resolution. Eades suggests this effective pixel size on film is mainly determined by the TEM operator's ability to focus the image using the usual binocular-screen system. However, the spatial resolution on the film can be improved to about 10 μm effective pixel size by very careful focusing of the TEM objective lens, such as can be accomplished with the aid of a post-column video-rate camera. Film digitization can be performed at up to 5000 dpi (5 μm pixel size) on a current flat-bed scanner. Thus, at this time, film offers similar, or with effort, slightly better resolution of detail in image contrast for the same microscope magnification. It is interesting to note that to resolve detail in image contrast down to a limit in weak-beam imaging of about 0.3 nm requires a microscope magnification of 2×10^5 based on a 25 μm pixel size on film or CCD chip. If, however, a 10 μm effective pixel size is achieved, the same weak-beam resolution limit can be reached at a microscope magnification of 10^5, increasing the recorded field of view by $4\times$.

For image contrast a grey scale depth of 8 bits (256 levels) is usually sufficient, certainly qualitatively, but probably quantitatively for comparison with image simulations to a similar bit depth. This depth is easily achieved by either digitized film or CCD camera. However, the CCD camera has a distinct advantage in linearity, especially at lower electron doses, which might be an advantage for shorter exposure times under weak-beam conditions. Another advantage of the CCD camera, useful primarily for quantitative diffraction data, is a much greater grey scale range (14 bit). An even larger grey scale range (16 bit) is available from an Image Plate system.

At present the most common size of CCD chip is 10^3 by 10^3 pixels, or about 2.5 cm square. This results in a field of view for the same microscope magnification about 13 times smaller in area than that of 8 cm $\times$ 10 cm film. Thus at relatively high magnification, 1–2 $\times 10^5$, the number of defects in a specimen area within the field of view can be quite restricted with a CCD camera. To form a montage of about 13 images would require considerable specimen and microscope stabilities. Of course, the size of CCD chips will increase in the future, and the cost per unit area is likely to decrease. Already 2×10^3 by 2×10^3 pixel chips are available.

For quantitative image-contrast measurements an energy filter, aligned to permit only elastically scattered electrons to form the image, is a distinct advantage. Little work has been done with energy filtered imaging of irradiation defects, but clearly a better resolution of very small defects (0.5 nm) will be achievable, and useful defect imaging in thicker foils will be possible. The use of an integrated post-filter CCD camera will provide the necessary digital recording and linear low-dose data required for quantitative analysis. The same system

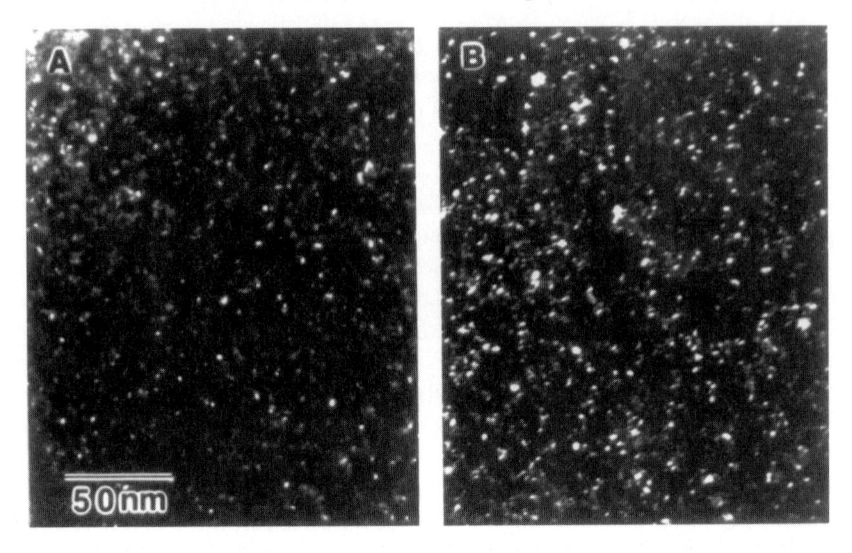

Figure 2.11. Caution required! Under some imaging conditions surface oxide can be confused with radiation damage. The micrographs show (*A*) small loops in ion-irradiated nickel, imaged under weak-beam conditions; and (*B*) surface oxide in unirradiated nickel, also under weak-beam conditions . At first sight these could be confused. In practice oxide can be recognized easily, both from its diffraction pattern, and the fact that the image moves when the objective lens is defocused. Micrographs courtesy of Dr J S Vetrano.

will be used to resolve elemental differences associated with structural defects in multicomponent materials with spatial resolution of several nanometres.

2.5 Comments on specimen preparation and image artefacts

The production of a 'representative' specimen, free from specimen artefacts, is one of the most difficult tasks in electron microscopy. Wherever possible a method should be chosen which minimizes the possibility of artefacts. Electrolytic or chemical polishing is generally preferable to ion-beam milling, but is not always feasible. Specimens produced by ion-beam milling always contain some type of ion-beam damage, often in the form of an amorphous surface layer. In unfavourable cases discrete damage clusters can be produced, which could be confused with the 'real' damage under investigation. If ion-beam milling cannot be avoided, milling artefacts can be minimized by reducing the ion energy and angle of incidence during the final stages of foil preparation, and by use of a liquid-nitrogen cold stage.

Even specimens produced by electropolishing are often contaminated with a fine-scale surface oxide. In unfavourable cases this could be confused with small clusters under some imaging conditions (see figure 2.11). The possibility

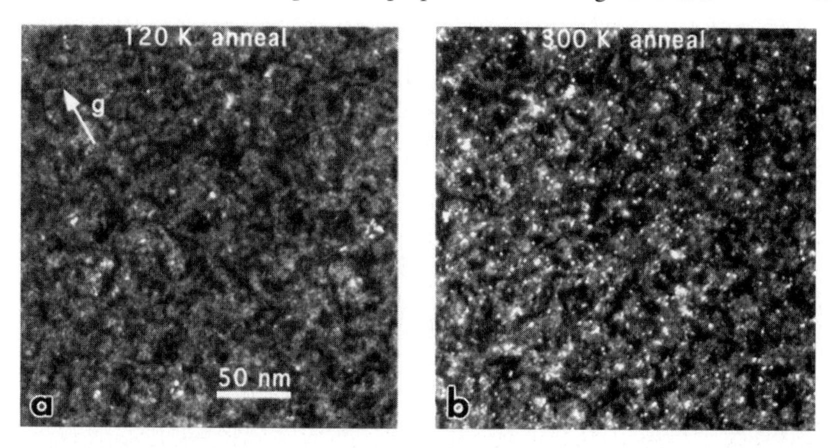

Figure 2.12. Another type of artefact. Micrograph (*a*) shows a weak-beam image of small defects in a copper specimen irradiated at 20 K with 600 keV Cu^+ ions and annealed to 120 K, while micrograph (*b*) shows the same foil area after warm-up to room temperature. The imaging conditions for both micrographs are similar: $\mathbf{g} = 200$ and $s_g = 2 \times 10^{-1}$ nm^{-1} and both images were recorded at 20 K after the anneals. A large number of extra defects (in fact, small stacking-fault tetrahedra) have appeared near the electron-exit surface of the foil. They are only seen in regions which were illuminated with electrons at low temperature. These observations are consistent with formation of vacancies by electron-beam (100 keV) sputtering of atoms from the back surface at 20 K, and subsequent vacancy clustering to form SFT upon annealing above the vacancy mobility stage III (250 K). From Kirk *et al* (2000).

that contaminants such as hydrogen may be introduced into the specimen during electropolishing must also be considered. Although it is usually possible to distinguish surface oxide from damage clusters, the presence of a surface oxide may well mask the smallest defects so that resolution is compromised. It is worth spending a considerable time in optimizing the polishing conditions to minimize this problem. Considerable care should also be taken in cleaning the specimen. Direct transfer into a cold washing solution, before switching off the current, often works well. Plasma cleaning to remove hydrocarbon contamination of the surface prior to microscopy, and the use of an anti-contaminator during microscopy, may also prove worthwhile. It should be remembered, however, that some discrete surface features are often an advantage, since they act as markers to locate the area of interest in contrast experiments. Foil bending can sometimes be avoided by preparing specimens with steep wedge angles, which can often be achieved by allowing the polishing to continue for several seconds after a hole first appears. Alternatively, the thinning process may be stopped just prior to perforation.

The objective of the polishing process is to produce a thin, flat foil with a smooth, relatively clean, surface. Even if this is achieved, other types of artefacts can occur. For example, dislocation loops near the final foil surface may glide to that surface under the influence of surface image forces and so be lost (see the discussion in section 5.1.3). At higher microscope operating voltages, the electron beam may produce knock-on damage. In non-metals the electron beam may cause ionization damage, which will be more severe at lower operating voltages. Even sub-threshold electron irradiations may result in sputtering from the bottom surface of the foil, with subsequent vacancy injection and possible clustering (see figure 2.12). More severe problems arise in *in situ* irradiations of thin foils, where the surfaces can act as sinks for migrating point-defects so that what occurs may not be typical of the bulk. In ion irradiations of thin foils, surface effects may sometimes be dominant—resulting in surface craters and other severe modifications to the damage structures. Of course, in semiconductor doping the presence of the surface is the 'real' situation, and surface effects can even be used to advantage in some situations. Examples of such effects will be found in various places in this book. With care, it is usually possible to recognize the presence of thin-foil artefacts and it is often possible to avoid them or otherwise take them into account.

Particular precautions to avoid artefacts may be required for *in situ* experiments carried out at low and high temperatures. These are discussed in chapter 9.

Chapter 3

Analysis of small centres of strain: the determination of loop morphologies

This chapter is concerned with the analysis of the geometry of small point-defect clusters such as dislocation loops and stacking-fault tetrahedra (SFT) which are imaged in diffraction contrast due to their elastic displacement fields. The defect structure—dislocation loop geometry in the present case—can determine such material properties as mechanical hardening by pinning of gliding dislocations. Knowledge of the defect structure can also give evidence as to how these defects were produced and may suggest ways to encourage or suppress their formation. In the present chapter, we are concerned with methods for determining the Burgers vectors, habit-planes, shapes and detailed morphologies of dislocation loops. We postpone discussions of how to determine the vacancy or interstitial nature of dislocation loops, and how to obtain quantitative measures of cluster sizes and number densities to chapters 4 and 5. We shall discuss two approaches for determining loop geometries. The Burgers vectors and habit-planes of dislocation loops of diameter smaller than about 10 nm are often best analysed using the black–white contrast method. This technique makes use of the characteristic image symmetries shown by clusters of different types when imaged under strong two-beam diffraction conditions. Black–white contrast is sensitive to the long-range strain fields of the clusters, and so gives relatively little information on features such as loop shapes. The second approach we shall discuss is weak-beam microscopy. Weak-beam microscopy is largely complementary to black–white analysis. It is sensitive to local strain fields and so is often useful for analysing clusters of complex shape or geometry, especially clusters larger than about 5 nm. It may be the only way to image and analyse successfully very small clusters lying close to the foil centre. The application of these methods will now be discussed in turn.

3.1 Black–white contrast analysis

Black–white contrast analysis may be used to analyse small centres of strain located within about ξ_g (where ξ_g is the extinction distance) of either foil surface (within the so-called layer structure, see section 3.1.1 and figure 3.2). Black–white images are obtained under dynamical two-beam conditions with the deviation parameter s_g zero or very small. Bright-field or dark-field images may be taken, although dark-field images are often preferred since these usually show higher contrast. Typical black–white images consist of two adjoining contrast lobes, one black and one white, although more complex contrast shapes consisting of several lobes are also possible. Any small centre of strain can give rise to black–white contrast. Examples of black–white images of small dislocation loops in copper have already been shown in figure 2.2. The symmetries of the black–white contrast shown by a given loop under different diffraction conditions depend on the loop Burgers vector and the loop habit-plane and so enable these to be determined, as will be discussed later. Examples of the black–white contrast of cobalt precipitates in copper with spherically-symmetric strain fields are shown in figure 3.1.

Much of this section is based on detailed comparisons between computer-simulated and experimental images of small loops. Image simulations may be made by numerical integration of the equations of the dynamical theory of electron diffraction. There are several physically-equivalent formulations: scattering by defects can be described in terms of plane waves (the Howie–Whelan formulation; Howie and Whelan 1961), Bloch waves (Wilkens 1964) or modified Bloch waves (Wilkens 1966). Usually the so-called column approximation is made. Since the scattering angle $2\theta_B$ for electrons is small, it is assumed that the intensity at a given point on the electron-exit surface of the specimen is determined by diffraction within a column of diameter about 1 nm above this point. A further approximation is to include just two diffracted beams, the forward-scattered beam and the diffracted beam $\boldsymbol{g}$. These approximations, as will be seen later, work well for small loops imaged under dynamical imaging conditions. They work less well for other diffraction conditions, for example weak-beam conditions. It is possible to develop the theory without the use of either approximation (see Howie and Basinski 1968), but although some calculations of loop contrast have been made using many-beam theory, to our knowledge none has been made without the use of the column approximation.

3.1.1 General properties of black–white contrast images

Most of the simulations shown here have been made using computer programs based on integration of the Howie–Whelan equations. However, for small clusters a better physical understanding can be achieved by considering the Bloch-wave approach. In dynamical two-beam diffraction conditions, just two Bloch waves have large amplitudes. Crystal imperfections cause a redistribution of amplitude

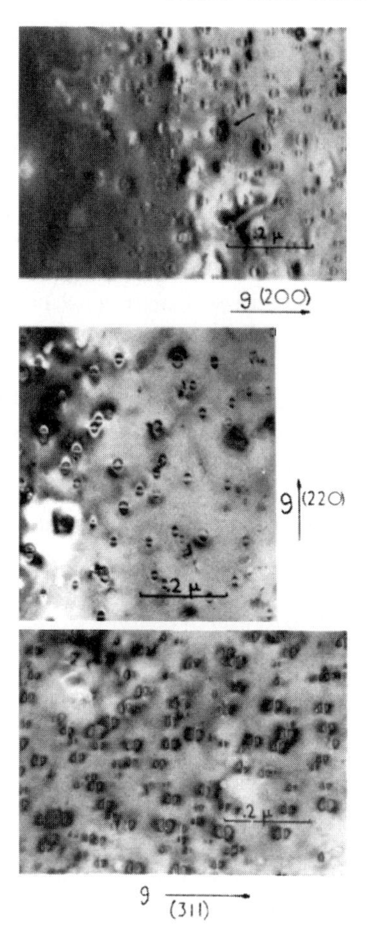

Figure 3.1. Small cobalt precipitates in copper, imaged in three different reflections as shown. The precipitates show double-arc contrast, in some cases with black–white contrast lobes. The $\boldsymbol{l}$ vectors are parallel to the diffraction vector $\boldsymbol{g}$ in each micrograph, consistent with a spherically-symmetric strain field. Taken from the work of Ashby and Brown (1963).

between these two Bloch waves. For defect clusters which are small in relation to the effective extinction distance ξ_g^{eff}, significant scattering between Bloch waves occurs only close to the cluster when $\mathrm{d}\boldsymbol{R}/\mathrm{d}z$, the z derivative of the defect displacement field $\boldsymbol{R}$, is of appreciable magnitude. In this case (called the kinematical scattering of Bloch waves) the Bloch wave equations can be integrated directly. Details of the theory are beyond the scope of this book. This approach was used by Hirsch (1978) to show analytically several of the general properties exhibited by black–white contrast images, including the following ones.

- Anomalous absorption makes defect clusters most strongly visible near the top and bottom of the foil and have weak visibility at the foil centre. The argument to show this is similar to that given by Hashimoto *et al* (1962) to explain the contrast of inclined stacking faults. In the usual notation, Bloch wave (1) has intensity maxima between crystal planes, while Bloch wave (2) has intensity maxima centred on the planes. Inelastic scattering processes will therefore be greater for wave (2) than for wave (1), and so wave (2) will be attenuated more quickly than wave (1). Consider defect clusters in a foil sufficiently thick that in regions of perfect crystal, wave (2) is completely absorbed at the bottom of the foil. For a defect located close to the top surface of the foil, where wave (2) still has appreciable intensity, interband scattering from wave (2) enhances wave (1). For a defect close to the bottom of the foil, interband scattering occurs from wave (1) to wave (2), which will not be much absorbed by the time it reaches the bottom of the foil. Both processes lead to high contrast. However, because of the high absorption of wave (2), neither process produces much contrast for a defect located near the centre of the foil. Only wave (1) reaches such a defect with appreciable intensity; interband scattering occurs from wave (1) into wave (2), but wave (2) is damped by the time it reaches the bottom of the foil.
- For dislocations, dislocation loops or other centres of strain, where the sign of $\beta = \boldsymbol{g} \cdot \mathrm{d}\boldsymbol{R}/\mathrm{d}z$ (where $\boldsymbol{R}$ is the defect displacement field, and z the depth of the defect in the foil) changes from one side of the defect (x) to the other, the contrast will appear bright on one side of the defect, and dark on the other. This is black–white contrast.
- For dislocation loops or other centres of strain where $\beta = \boldsymbol{g} \cdot \mathrm{d}\boldsymbol{R}/\mathrm{d}z$ is an odd function of z, the contrast in the dark field is the same for a defect near the top of the foil, and for a similar distance from the bottom. In a bright field the contrast is opposite for a defect near the top of the foil and for a similar distance from the bottom.
- For dislocation loops and other small centres of strain, beating between the Bloch waves results in the contrast switching sign at depths $\frac{1}{4}\xi_g$, $\frac{3}{4}\xi_g$ and $\frac{5}{4}\xi_g$. Loops further from the surface do not show black–white contrast because of anomalous absorption, as discussed previously. This is the well-known layer structure for small loops (Rühle *et al* 1965). There will be a corresponding layer structure at the bottom surface. In practice the layer structure may be modified by surface relaxation effects and the transition from first to second layer occurs typically at $\sim\frac{1}{3}\xi_g$ for small loops (see, e.g., Wilkens 1978). The layer structure is shown schematically in figure 3.2. The depth dependence of the black–white contrast is the basis of a method for determining the interstitial or vacancy nature of small clusters. This is done by an analysis of the direction $\boldsymbol{l}$ of black–white streaking with respect to the diffraction vector $\boldsymbol{g}$ (see later), together with measurements of the depth of the loop in the foil. This is discussed in detail in chapter 4.

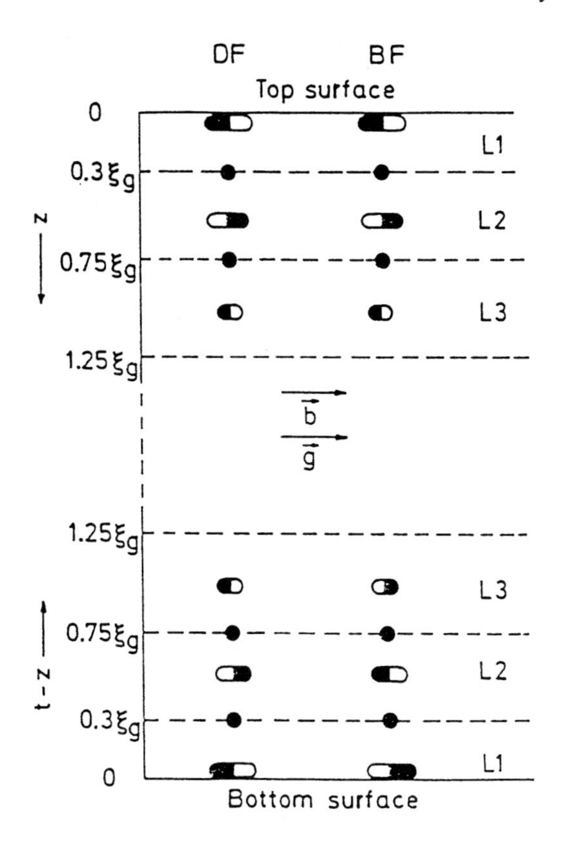

Figure 3.2. Depth oscillations of the black–white contrast vector $\boldsymbol{l}$ for a small vacancy loop, showing the so-called layer structure. The thickness of layer L1 depends on surface relaxation effects—see the text. For the same diffraction vector $\boldsymbol{g}$ the direction of $\boldsymbol{l}$ for an interstitial loop would be reversed. Reversing $\boldsymbol{g}$ also reverses $\boldsymbol{l}$. This contrast behaviour may be used as the basis for loop nature determination, as discussed in chapter 4.

3.1.2 Burgers vector determination by l-vector analysis

For small loops, the $\boldsymbol{g} \cdot \boldsymbol{b} = 0$ invisibility criterion, which is used to determine the Burgers vectors of line dislocations, may be of limited utility. Loops with $\boldsymbol{g} \cdot \boldsymbol{b} = 0$ are often not invisible when imaged under dynamical two-beam conditions. They may show characteristic contrast under some conditions—for example 'butterfly images'—but it is possible to confuse some $\boldsymbol{g} \cdot \boldsymbol{b} = 0$ images with some $\boldsymbol{g} \cdot \boldsymbol{b} \neq 0$ images. This is discussed in section 3.1.3. Under weak-beam imaging conditions, loops usually show invisibility or very weak contrast when $\boldsymbol{g} \cdot \boldsymbol{b} = 0$. However, small loops with $\boldsymbol{g} \cdot \boldsymbol{b} \neq 0$ may also show very weak contrast under weak-beam imaging conditions. This can be confused with true invisibility (see section 3.2).

In either case, a contrast experiment using various diffraction vectors would be required for a full analysis. This has led some researchers to advocate '$\boldsymbol{l}$-vector analysis' as an alternative and easier means of determining the Burgers vectors of small loops. The method relies on examining the changes in directions of black–white streaking with the operating diffraction vector $\boldsymbol{g}$. The black–white streaking direction is defined by the so-called $\boldsymbol{l}$-vector. For a given black–white contrast figure, the $\boldsymbol{l}$-vector is defined to run from the centre of the black contrast lobe to the centre of the white contrast lobe on a positive print. For dislocation loops, the direction of $\boldsymbol{l}$ in many situations is parallel to $\boldsymbol{b}$, or its projection $\boldsymbol{b}_\mathrm{p}$ on the image plane. For an evaluation of $\boldsymbol{b}$, it is then necessary only to evaluate the direction of $\boldsymbol{l}$ for at least two images taken at different foil orientations.

$\boldsymbol{l}$-vector analysis immediately allows dislocation loops to be distinguished from small misfitting precipitates or SFT. For defects with symmetrical strain fields, such as the spherical precipitates shown in figure 3.1, the $\boldsymbol{l}$-vector is found to run approximately parallel to $\boldsymbol{g}$. Other defects with approximately spherically-symmetrical strain fields include SFT and small bubbles (see chapters 4 and 6). For dislocation loops, however, the direction of the $\boldsymbol{l}$-vector is usually tied fairly closely to the Burgers vector $\boldsymbol{b}$ or its projection on the image plane $\boldsymbol{b}_\mathrm{p}$. If a series of micrographs such as those in figure 2.2 or figure 3.1 is recorded with different diffraction vectors $\boldsymbol{g}$, the behaviour of $\boldsymbol{l}$ with respect to $\boldsymbol{g}$ allows conclusions to be drawn about the strain field of the defects. It can be seen in figure 2.2 that the $\boldsymbol{l}$-vectors do not rotate to follow $\boldsymbol{g}$ from one micrograph to another. These defects are therefore dislocation loops.

Examination of the directions of the $\boldsymbol{l}$-vectors of the loops arrowed in figure 2.2 show that they lie close to the $[11\bar{1}]$ direction, showing that these loops are of the faulted Frank type with $\boldsymbol{b} = \frac{1}{3}[11\bar{1}]$. Note that in the first three images (*a*)–(*c*) the line of no contrast between black and white lobes of the images of these edge-on loops is straight and lies parallel to the projection of the $(11\bar{1})$ habit-plane. This is a feature characteristic of edge-on Frank loops. Other characteristic image features of loops are described later.

$\boldsymbol{l}$-vector analysis generally works well for edge loops if the angle between $\boldsymbol{g}$ and $\boldsymbol{b}$ is small. However, as we shall see later, the simple $\boldsymbol{l}$-vector approach may break down when the angle between $\boldsymbol{g}$ and $\boldsymbol{b}$ or $\boldsymbol{b}_\mathrm{p}$ is large, or when non-edge loops are present. Difficulties in distinguishing between loops and small precipitates may also occur if the strain field of the precipitate is anisotropic—which may be the case if the precipitate is not spherical or the material has high elastic anisotropy. $\boldsymbol{l}$-vector analysis in these cases is not sufficient for a proper analysis. Detailed image matching is necessary, as discussed later. It may, however, be worth mentioning the utility of carrying out different metallurgical treatments (for example, thermal aging) to grow precipitates to sizes more amenable for analysis and the use of control experiments (using, for example, unirradiated specimens). Such experiments may enable one to recognize the presence of small precipitates rather easily.

3.1.3 Image simulations

It has already been mentioned that a widely adopted approach to determining the Burgers vectors and habit-planes of loops is to match experimental images with computer-generated images. The image simulations are made using computer programs which integrate the equations of the dynamical theory of electron diffraction. Unfortunately, few simulation packages are now available for this purpose. The most extensive simulations have been made using the Harwell code, which is based on the two-beam Howie–Whelan equations[1]. In many situations it may be sufficient to make the comparison with image simulations published in the literature. Some of the more important papers with such simulations are summarized in table 3.1.

Note the following points.

- The majority of the simulations in the literature have used two-beam dynamical theory in the column approximation. These approximations appear to work well provided that the Bragg condition is satisfied (i.e. $s_g = 0$). There have been few simulations of loop contrast under kinematical or weak-beam diffraction conditions, where these approximations are less good.
- Most of the simulations have been for loops located within the first few depth layers, usually the first depth layer L1, and well away from layer boundaries. Eyre *et al* (1977a) have, however, reported simulations of loops within the transition zones between layers and the main conclusions of this work are described in section 3.1.3.6.
- The calculations require knowledge of the displacement field of a loop in a thin foil. If the loop lies close to the foil surface, and if this surface is not constrained by say an oxide layer, then the strain field of a loop in a *semi-infinite* medium is required. Most authors, however, have used an expression for the strain field due to a small loop in an *infinite* medium derived from linear elasticity theory and so neglect the effect of the nearby surface. The most extensive simulations, those of Eyre and co-workers, are of this type. Other authors (e.g. Häussermann *et al* 1972, Sykes *et al* 1981) have used the infinitesimal loop approximation, in most cases also ignoring any surface effects. The effect of a free surface has been examined explicitly by Ohr (1977).
- Elastic anisotropy has not been included in most cases. An exception is the work of Ohr (1979). His conclusions are summarized in section 3.1.3.7.
- As regards the overall symmetries of the black–white contrast figures, the different approaches seem to give similar results and good correspondence with experiments. In particular, the simulations of Eyre and co-workers, which ignore both the free surface and elastic anisotropy, have proven very successful for analysing loops in molybdenum, ruthenium and copper. The

[1] This code is not available commercially, but is available by contacting either of the authors.

Table 3.1.

Material	Structure	Loop geometry Burgers vectors $\boldsymbol{b}$ Diffraction vectors $\boldsymbol{g}$ Foil normal z	Comments	References
Molybdenum	bcc	Circular edge loops $\boldsymbol{b} = \frac{1}{2}\langle 110\rangle, \langle 100\rangle \frac{1}{2}\langle 111\rangle$ $z = [0\bar{1}1]$; $\boldsymbol{g} = 01\bar{1}, 21\bar{1}, 200, 2\bar{1}1$ $z = [001]$, $\boldsymbol{g} = \bar{1}10, 020, 110, 200$	Isotropic elasticity theory, extensive simulations	Maher *et al* (1971) Eyre *et al* (1977a, b)
Molybdenum	bcc	Circular loops with shear components $\boldsymbol{b} = \frac{1}{2}\langle 110\rangle, \langle 100\rangle$ $z = [\bar{0}11]$; $\boldsymbol{g} = 01\bar{1}, 21\bar{1}, 200, 2\bar{1}1$ $z = [001]$, $\boldsymbol{g} = \bar{1}10, 020, 110, 200$	Isotropic elasticity theory, extensive simulations	Holmes *et al* (1979) English *et al* (1977, 1980)
Copper	fcc	Infinitesimal loops	Isotropic elasticity theory	Wilkens and Rühle (1972) Häussermann *et al* (1972)
Copper	fcc	Hexagonal Frank loops $\boldsymbol{b} = \frac{1}{3}\langle 111\rangle$ SFT	Angular dislocation method, isotropic elasticity theory	Saldin *et al* (1979)
Copper, nickel, niobium	fcc bcc	Circular Frank loops $\boldsymbol{b} = \frac{1}{3}[1\bar{1}1]$ Circular edge loops $\boldsymbol{b} = \frac{1}{2}[1\bar{1}1]$ $z = \bar{1}10$ $\boldsymbol{g} = 002$	Anisotropic elasticity theory Calculation for one condition only	Ohr (1979)
Ruthenium	hcp	Circular loops with shear components $\boldsymbol{b} = \frac{1}{2}\langle 10\bar{1}0\rangle$ $\frac{1}{3}\langle 11\bar{2}0\rangle$ $\frac{1}{6}\langle 20\bar{2}3\rangle$ $\frac{1}{2}[0001]$ $z = [0001], [11\bar{2}3], [12\bar{3}0]$ $\boldsymbol{g} = \{02\bar{2}0\}, \{11\bar{2}0\}, \{11\bar{2}2\}$	Isotropic elasticity theory	Phythian *et al* (1987)
Aluminium, tungsten	fcc bcc	Infinitesimal loops Large number of loop crystallographies, beam directions, and diffraction vectors	Isotropic elasticity theory Extensive catalogue of images	Sykes *et al* (1981)

infinitesimal loop approximation gives good agreement for the overall shape of the contrast figures, but breaks down close to the centre of the contrast figure, and so does not reproduce correctly the contrast details of the black–white interface.

- Further complications may arise if the defect geometry is itself complicated.

For example, black–white interfaces of defects in copper sometimes display a characteristic angular fine structure. This has been interpreted by Wilson and Hirsch (1972) as due to a partial loop dissociation. This interpretation has subsequently been confirmed in weak-beam studies by Jenkins (1974). This is discussed further in section 3.2.

Before going on to look at some specific examples of image matching it is useful to look at some further general features of black–white images, which have emerged from the simulations. We shall examine the influence of factors such as Burgers vector $\boldsymbol{b}$, loop normal $\boldsymbol{n}$, diffraction vector $\boldsymbol{g}$, deviation parameter s_g, image plane normal $\boldsymbol{z}$, loop depth and size and foil thickness.

3.1.3.1 *The four basic black–white contrast types for edge loops*

Image matching is facilitated by the occurrence for edge loops within the layer structure of four distinct black–white contrast types. These contrast types depend on the magnitude of $|\boldsymbol{g}\cdot\boldsymbol{b}|$ and the dislocation loop orientation. They are illustrated in figure 3.3 which shows simulated images of edge loops in the bcc metal, molybdenum. Shown in this figure is the expected contrast of edge loops with Burgers vectors of the types $\frac{1}{2}\langle 011\rangle$, $\frac{1}{2}\langle 111\rangle$ and $\langle 100\rangle$ for the four reflections available at $\boldsymbol{z} = [011]$: (*a*) $\boldsymbol{g} = 01\bar{1}$; (*b*) $\boldsymbol{g} = 21\bar{1}$; (*c*) $\boldsymbol{g} = 200$; and (*d*) $\boldsymbol{g} = 2\bar{1}1$. The corresponding values of $|\boldsymbol{g}\cdot\boldsymbol{b}|$ for these simulations are shown in table 3.2, and the crystallographic relationships can be visualized from the stereograph in figure 3.4. Only dark-field images are shown, for loops situated in the first layer at the bottom of the foil. In each case the contrast in the bright-field image was similar to that in the dark field.

The four contrast types have been classified by Eyre *et al* (1977a) as follows.

- *Type 1.* $|\boldsymbol{g}\cdot\boldsymbol{b}| = 0$, and the acute angle between the loop normal $\boldsymbol{n}$ and the beam direction $\boldsymbol{z}$ is less than about 45° (e.g. the top row of figure 3.3[2]). The image is weak and probably invisible except for higher-order reflections such as {211}. If visible, a simple black–white image with $\boldsymbol{l}$ parallel to $\boldsymbol{g}$ is predicted.
- *Type 2.* $|\boldsymbol{g}\cdot\boldsymbol{b}| = 0$, and the acute angle between the loop normal $\boldsymbol{n}$ and the beam direction $\boldsymbol{z}$ is equal or close to 90° (e.g. figure 3.3, row 3, column (*c*)). A symmetric 'butterfly' image, consisting of one small and two large pairs of alternating black–white lobes, is seen. The small black–white lobe is aligned parallel to $\boldsymbol{g}$.
- *Type 3.* $0 < |\boldsymbol{g}\cdot\boldsymbol{b}| \le 1$ (e.g. figure 3.3, row 2, columns (*a*), (*b*) and (*d*)). This is a simple black–white lobe with no interface structure, independent of the angle between $\boldsymbol{n}$ and $\boldsymbol{z}$.

[2] Not all examples of the four contrast types seen in figure 3.3 are pointed out explicitly here. The reader should consult table 3.2 and figure 3.4, and verify that each image of figure 3.3 does indeed correspond to one of the four contrast types described, depending on the magnitude of $|\boldsymbol{g}\cdot\boldsymbol{b}|$ and the dislocation loop orientation.

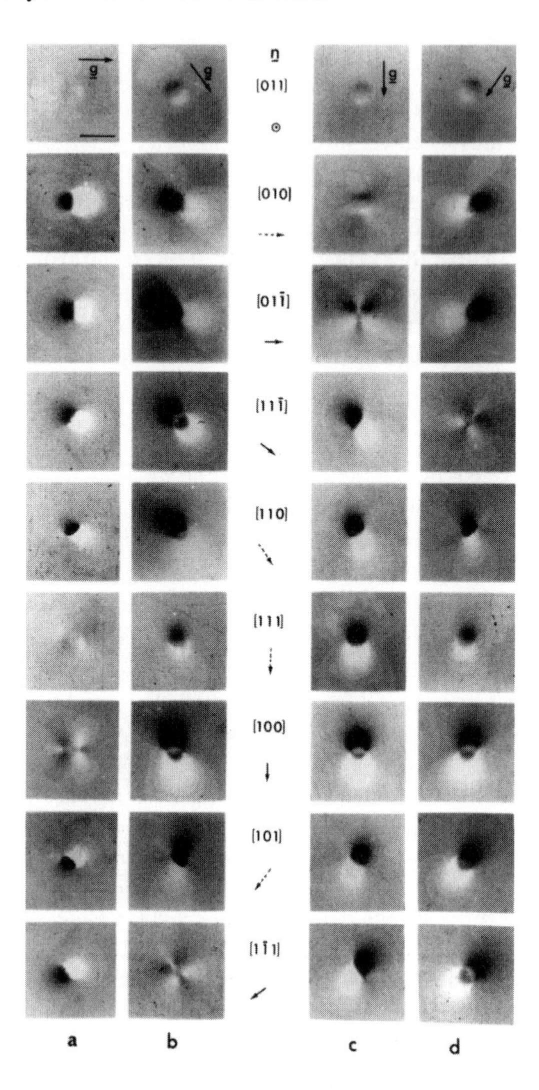

Figure 3.3. Simulated dark-field images of edge dislocation loops in molybdenum, with Burgers vectors of the types $\frac{1}{2}\langle 011\rangle$, $\frac{1}{2}\langle 111\rangle$ and $\langle 100\rangle$, taken from the work of Eyre *et al* (1977a). The images are all for a foil normal $z = 011$, and in each case $s_g = 0$. The loops lie in the first layer L1 at the bottom of the foil. The specific Burgers vector for each row is shown in square brackets at the centre of the row. The loop-plane normals (full arrows) or their projection on the image plane (broken arrows) are as shown and apply to each row of images. The four columns show simulated images for reflections (*a*) $\boldsymbol{g} = 01\bar{1}$, (*b*) $\boldsymbol{g} = 21\bar{1}$, (*c*) $\boldsymbol{g} = 200$ and (*d*) $\boldsymbol{g} = 2\bar{1}1$. The directions of these $\boldsymbol{g}$-vectors are shown in the images of the top row of the figure. Foil thickness $t = 5.25$, loop depth $d = 5.10$, loop radius 0.1, in units of ξ_{110}. Scale mark (at top left) 10 nm.

Table 3.2. $|\boldsymbol{g} \cdot \boldsymbol{b}|$ values for a [011] foil normal.

	$+\boldsymbol{g}$			
$+\boldsymbol{b}$	$[01\bar{1}]$	$[21\bar{1}]$	[200]	$[2\bar{1}1]$
[011]	0	0	0	0
[010]	1	1	0	1
$\frac{1}{2}[01\bar{1}]$	1	1	0	1
$\frac{1}{2}[11\bar{1}]$	1	2	1	0
$\frac{1}{2}[110]$	$\frac{1}{2}$	$\frac{3}{2}$	1	$\frac{1}{2}$
$\frac{1}{2}[111]$	0	1	1	1
[100]	0	2	2	2
$\frac{1}{2}[101]$	$\frac{1}{2}$	$\frac{1}{2}$	1	$\frac{3}{2}$
$\frac{1}{2}[1\bar{1}1]$	1	0	1	2

- *Type 4.* $|\boldsymbol{g} \cdot \boldsymbol{b}| > 1$ (e.g. figure 3.3, row 7, columns (*b*), (*c*) and (*d*)). This is again a black–white lobe but now the interface between the lobes has structure, for example a small white area of contrast within the dark lobe.

The general features of type 2 and type 3 images may be understood intuitively by considering the long-range displacement field of a small loop, following the argument of Sykes *et al* (1981). Diffraction theory shows that the quantity which enters the scattering equations is $\beta = \mathrm{d}(\boldsymbol{g} \cdot \boldsymbol{R})/\mathrm{d}z$, that is the image is sensitive only to displacements in the direction of $\boldsymbol{g}$. Figure 3.5(*a*) at the bottom left shows schematically the displacements $\boldsymbol{R}'$ at a constant distance from an edge-on interstitial loop in a plane intersecting the loop. In figure 3.5(*b*)–(*d*), these displacements are resolved in the direction of three possible diffraction vectors $\boldsymbol{g}$, that is the plots show $\boldsymbol{r}(\theta) = \boldsymbol{g} \cdot \boldsymbol{R}'(\theta)$. If $\boldsymbol{g} \cdot \boldsymbol{R}'(\theta)$ is positive a full line is plotted, if negative a broken line is given. Above the line diagrams are simulated loop images corresponding to these diffraction geometries. The correspondence between the plots of the displacement field and the simulated images is striking. Figure 3.5(*b*) also makes clear the origin of type 2 'butterfly' images when $\boldsymbol{g} \cdot \boldsymbol{b} = 0$. The small subsidiary lobes parallel to $\boldsymbol{g}$ arise from the inward collapse of material at the top and bottom of the circumference shown in figure 3.5(*a*). The fine structure of the black–white interface seen in the simulated image of figure 3.5(*d*) depends on the local displacement field close to the loop, and so type 4 images cannot be explained intuitively in this way.

Simulations have confirmed that edge dislocation loops in fcc and hcp materials also always show one of the four basic types of contrast described here, depending on $|\boldsymbol{g} \cdot \boldsymbol{b}|$, although there may be small variations in the detailed appearance of the contrast depending on the angle between $\boldsymbol{g}$ and $\boldsymbol{b}$ (Eyre *et al*

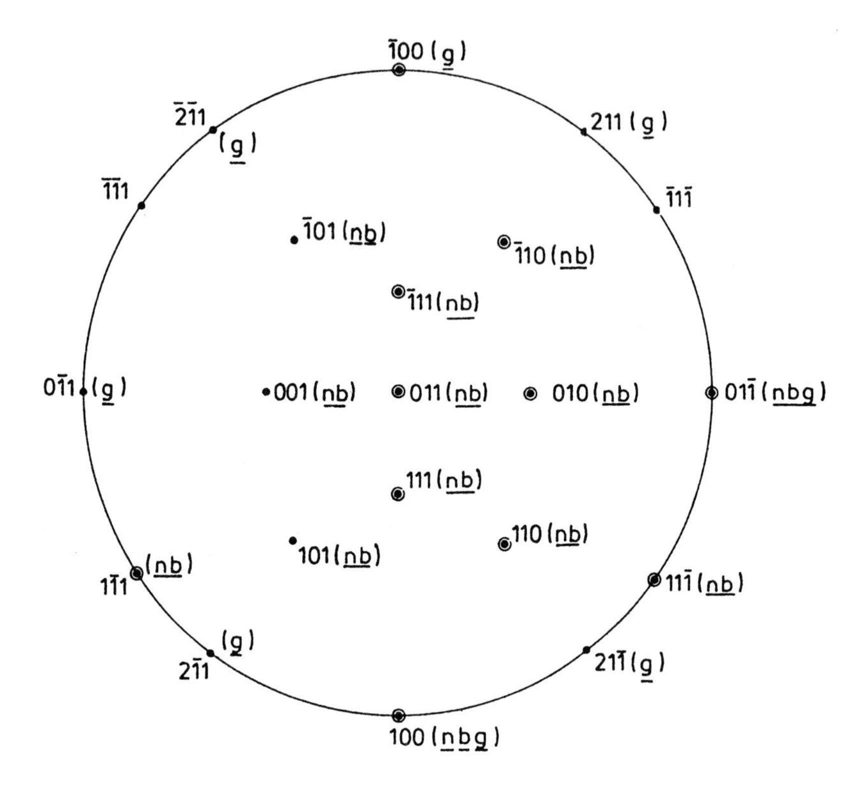

Figure 3.4. Stereographic projection illustrating the crystallographic relationships between the possible Burgers vectors $\boldsymbol{b}$ of edge dislocation loops $\boldsymbol{n}$ and the diffraction vectors $\boldsymbol{g}$ for a bcc crystal with $z = [011]$.

1977b, Saldin *et al* 1979, Phythian *et al* 1987). Examples of simulations of edge loops in an isotropic fcc material are shown in figure 3.6, which may be compared with the experimental images of loops in copper shown in figure 2.2. There is a reasonably good correpondence between the simulated images of Frank loops shown in figure 3.6(*e*)–(*h*) and the experimental images[3]. A comparison between simulated and experimental images of loops in hcp ruthenium is described in section 3.1.4.

3.1.3.2 Implications for $\boldsymbol{l}$*-vector analysis*

When the angle between $\boldsymbol{g}$ and $\boldsymbol{b}$ is small, the simulations shown in figure 3.3 show that the direction of $\boldsymbol{l}$ is indeed close to $\boldsymbol{b}$ or $\boldsymbol{b}_{\mathrm{p}}$. However, as the angle

[3] In these simulations for an isotropic material the $\boldsymbol{l}$-vector direction deviates rather more from the direction of $\boldsymbol{b}$ than is seen in practice. If elastic anisotropy is included in the simulation, $\boldsymbol{l}$ lies closer to $\boldsymbol{b}$, see section 3.1.3.7.

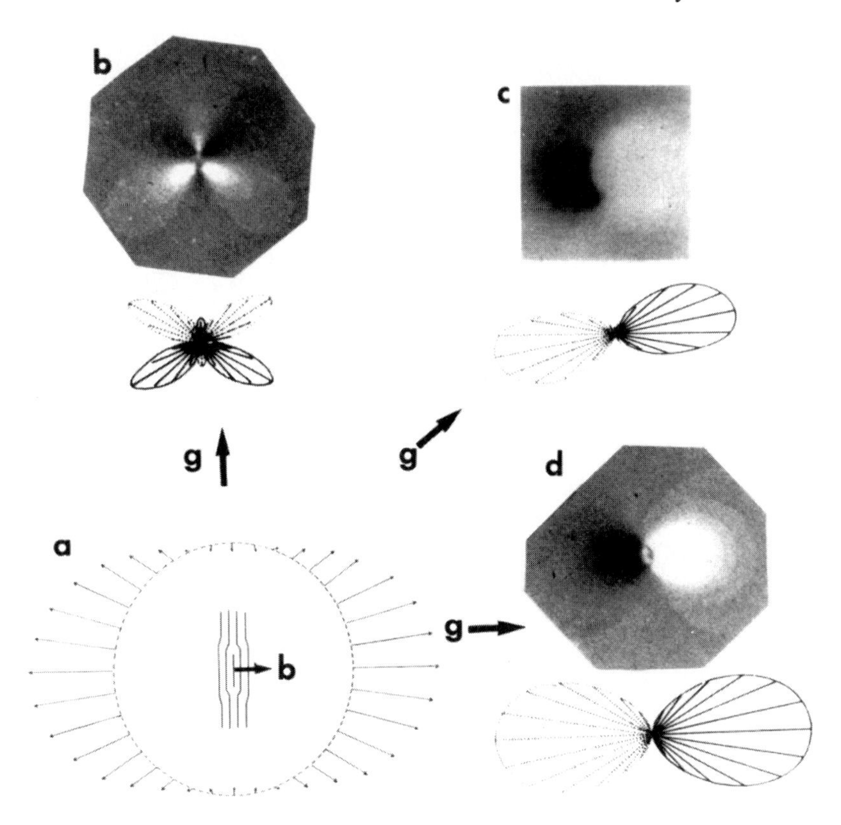

Figure 3.5. Figure to illustrate the relationship between the images and the loop displacement field. (*a*) Shows schematically the displacements at a constant distance from an edge-on interstitial loop in a plane intersecting the loop. In (*b*)–(*d*) these displacements are resolved in the direction of three possible diffraction vectors $\boldsymbol{g}$ corresponding to (*b*) $\boldsymbol{g} \cdot \boldsymbol{b} = 0$, (*c*) $\boldsymbol{g} \cdot \boldsymbol{b} = 1$, and (*d*) $\boldsymbol{g} \cdot \boldsymbol{b} = 2$. Above each line diagram is shown a simulated loop image corresponding to this diffraction geometry. Modified from figure 15 of Sykes *et al* (1981).

between $\boldsymbol{g}$ and $\boldsymbol{b}$ increases, the direction of $\boldsymbol{l}$ deviates increasingly from the direction of $\boldsymbol{b}$ or $\boldsymbol{b}_\mathrm{p}$ to reach values of 30° or more. Figure 3.7 demonstrates this explicitly. If the total number of possible Burgers-vector directions is large there may be no imaging orientations where $\boldsymbol{b}$ can be deduced unambiguously from a knowledge of $\boldsymbol{l}$. This is the case, for example, in bcc materials such as molybdenum where there are 13 different permutations of the most likely Burgers vectors $\boldsymbol{b} = \frac{1}{2}\langle 111\rangle$, $\langle 100\rangle$ and $\frac{1}{2}\langle 110\rangle$. Simple $\boldsymbol{l}$-vector analysis may therefore fail even in the simple case of edge loops in an isotropic elastic material.

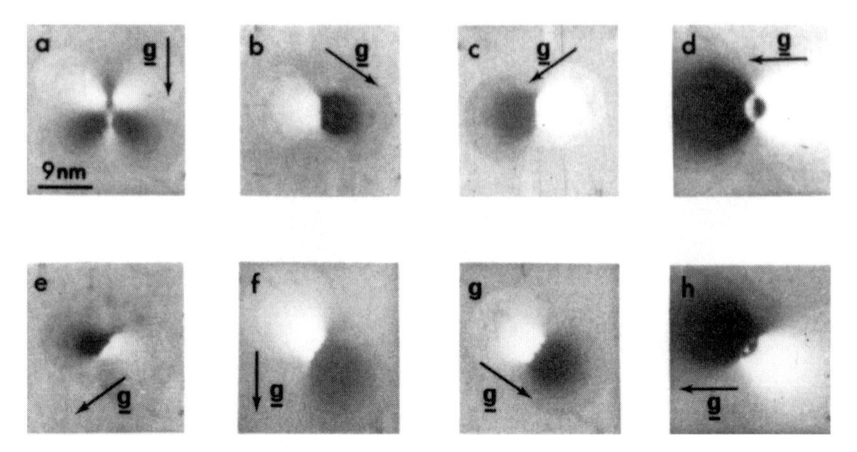

Figure 3.6. Computer-simulated, dark-field images of edge dislocation loops in an isotropic fcc crystal with $z = [011]$ and $s_g = 0$. Simulations (*a*)–(*d*) are for a perfect loop with $\boldsymbol{b} = \frac{1}{2}[01\bar{1}]$, whilst simulations (*e*)–(*h*) are for a Frank loop with $\boldsymbol{b} = \frac{1}{3}[11\bar{1}]$. From Eyre *et al* (1977b). (*a*) $\boldsymbol{g} = 200$, $\boldsymbol{g} \cdot \boldsymbol{b} = 0$; (*b*) $\boldsymbol{g} = 11\bar{1}$, $\boldsymbol{g} \cdot \boldsymbol{b} = 1$; (*c*) $\boldsymbol{g} = 1\bar{1}1$, $\boldsymbol{g} \cdot \boldsymbol{b} = 1$; (*d*) $\boldsymbol{g} = 0\bar{2}2$, $\boldsymbol{g} \cdot \boldsymbol{b} = 2$; (*e*) $\boldsymbol{g} = 1\bar{1}1$, $\boldsymbol{g} \cdot \boldsymbol{b} = \frac{1}{3}$; (*f*) $\boldsymbol{g} = 200$, $\boldsymbol{g} \cdot \boldsymbol{b} = \frac{2}{3}$; (*g*) $\boldsymbol{g} = 11\bar{1}$, $\boldsymbol{g} \cdot \boldsymbol{b} = 1$; (*h*) $\boldsymbol{g} = 0\bar{2}2$, $\boldsymbol{g} \cdot \boldsymbol{b} = \frac{4}{3}$. Foil thickness $t = 5.25$, loop depth $d = 5.10$, loop radius 0.1, in units of ξ_{111}. For Cu at 100 keV, $\xi_{111} = 24.3$ nm.

3.1.3.3 Orientation dependence of the contrast

Image simulations of defects are most easily calculated with beam directions down the major crystal zone axes. However, experimental two-beam diffraction conditions are, of course, best established for beam directions 8–10° away from a zone axis. For this small degree of crystal tilt the simulated defect images are found to be virtually indistinguishable from those without tilt, thus justifying the use of zone axis orientations in image simulations.

3.1.3.4 The effect of changing the loop size

Eyre *et al* (1977a) have confirmed that the characteristic features of type 2 and type 4 images are preserved for loops at the centre of layer L1 when the loop radius is reduced from $0.1\xi_{110}$ (2.3 nm) to $0.05\xi_{110}$ (1.2 nm). At this smaller size, loop identification is likely to be limited by specimen and signal-to-noise considerations and the microscope resolution.

They also confirmed that loops showing type 3 contrast did not develop interfacial structure which could be confused with type 4 contrast when the loop radius was increased to $0.25\xi_{110}$, when the loop is comparable in size to the layer thickness.

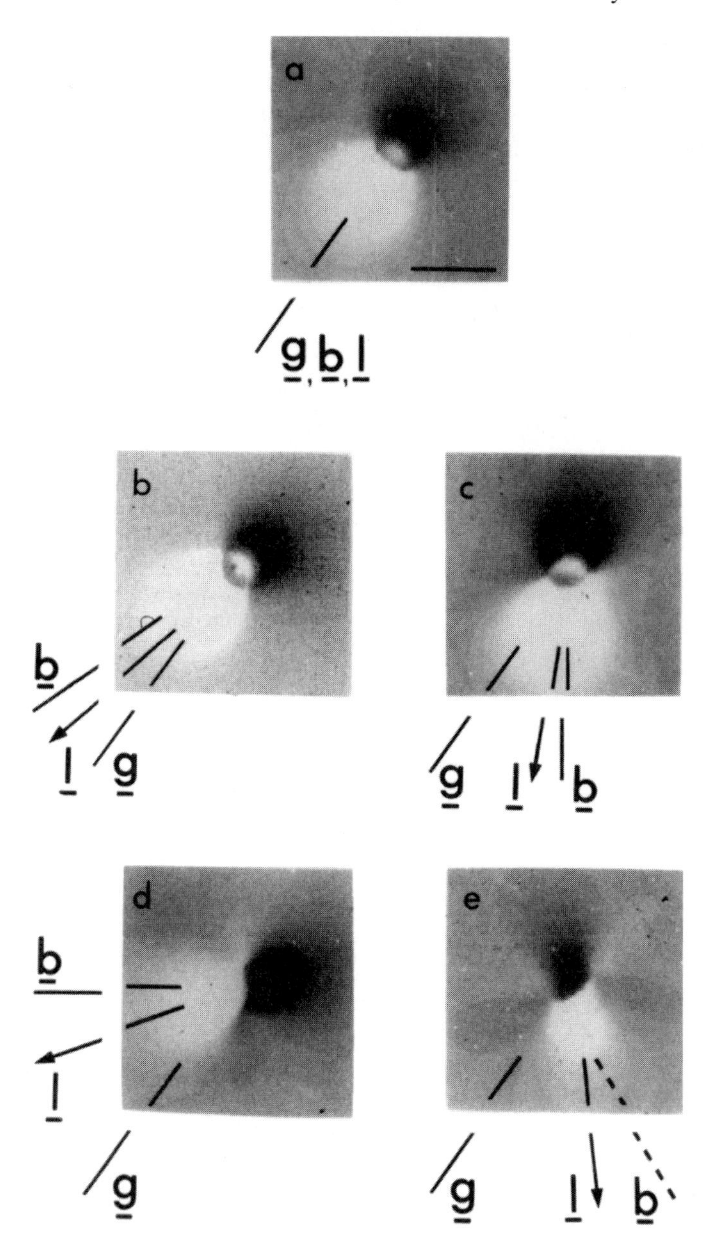

Figure 3.7. Simulated dark-field images of edge loops in molybdenum ($z = [011]$, $s_g = 0$, $g = \pm\bar{2}\bar{1}1$), showing the relationship between $\boldsymbol{l}$ and $\boldsymbol{b}$ (or $\boldsymbol{b}_\mathrm{p}$) as the angle between $\boldsymbol{g}$ and $\boldsymbol{b}$ increases. (*a*) $\boldsymbol{b} = \frac{1}{2}[101]$; (*b*) $\boldsymbol{b} = \frac{1}{2}[1\bar{1}1]$; (*c*) $\boldsymbol{b} = \frac{1}{2}[100]$; (*d*) $\boldsymbol{b} = \frac{1}{2}[0\bar{1}1]$; (*e*) $\boldsymbol{b} = \frac{1}{2}[110]$. Foil thickness $t = 5.25$, loop depth $d = 5.10$, loop radius 0.1, in units of ξ_{110}. For Cu at 100 keV, $\xi_{110} = 23.2$ nm. Scale mark 10 nm. From Eyre *et al* (1977b).

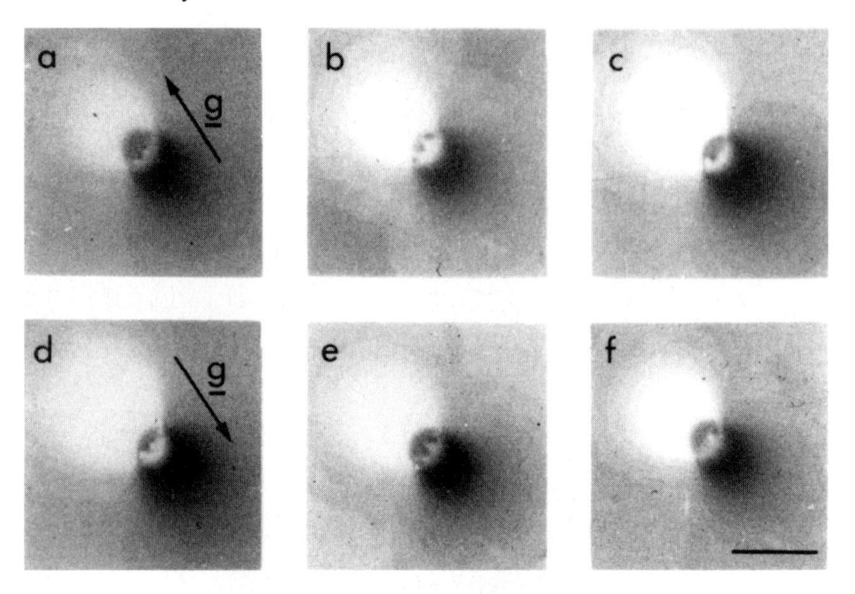

Figure 3.8. Bright-field ((*a*)–(*c*)) and dark-field ((*d*)–(*e*)) images of edge dislocation loop in molybdenum, showing the calculated variation in type 4 images as a function of the foil thickness. Loop parameters: $\boldsymbol{z} = [011]$, $s_g = 0$, $\boldsymbol{b} = \frac{1}{2}[11\bar{1}]$, $\boldsymbol{g} = \pm\bar{2}\bar{1}1$ (so that $|\boldsymbol{g} \cdot \boldsymbol{b}| = 2$) and in each case the loop is located close to the electron exit surface and in layer L1. Scale mark 10 nm. In (*a*) and (*d*) $t = 3.0\xi_{211}$ $(= 5.3\xi_{110})$; in (*b*) and (*e*) $t = 3.25\xi_{211}$ $(= 5.8\xi_{110})$; and in (*c*) and (*f*) $t = 3.5\xi_{211}$ $(= 6.2\xi_{110})$. From Eyre *et al* (1977b).

3.1.3.5 The effect of the foil thickness and defect depth

The calculated variation in type 4 contrast as a function of foil thickness, with other imaging parameters held constant, is shown in figure 3.8. It can be seen that the essential features of the images are preserved, implying that precise knowledge of the foil thickness is not necessary for successful defect identification.

The variation of image contrast as a function of loop depth is more complicated. For defects located at the centres of the depth layers, the black and white image lobes alternate in successive layers as indicated schematically in figure 3.2, and shown in figure 3.9, but in this case the essential features of the images also remain unchanged for all image types. In practice, loops lying in layers 2 and 3 usually show weaker contrast than first-layer loops (see section 4.2) although this does not seem to be shown in the calculations of figure 3.9, probably because of intensity renormalization.

Loops situated close to the transition zones between layers present a particular problem, and are considered next.

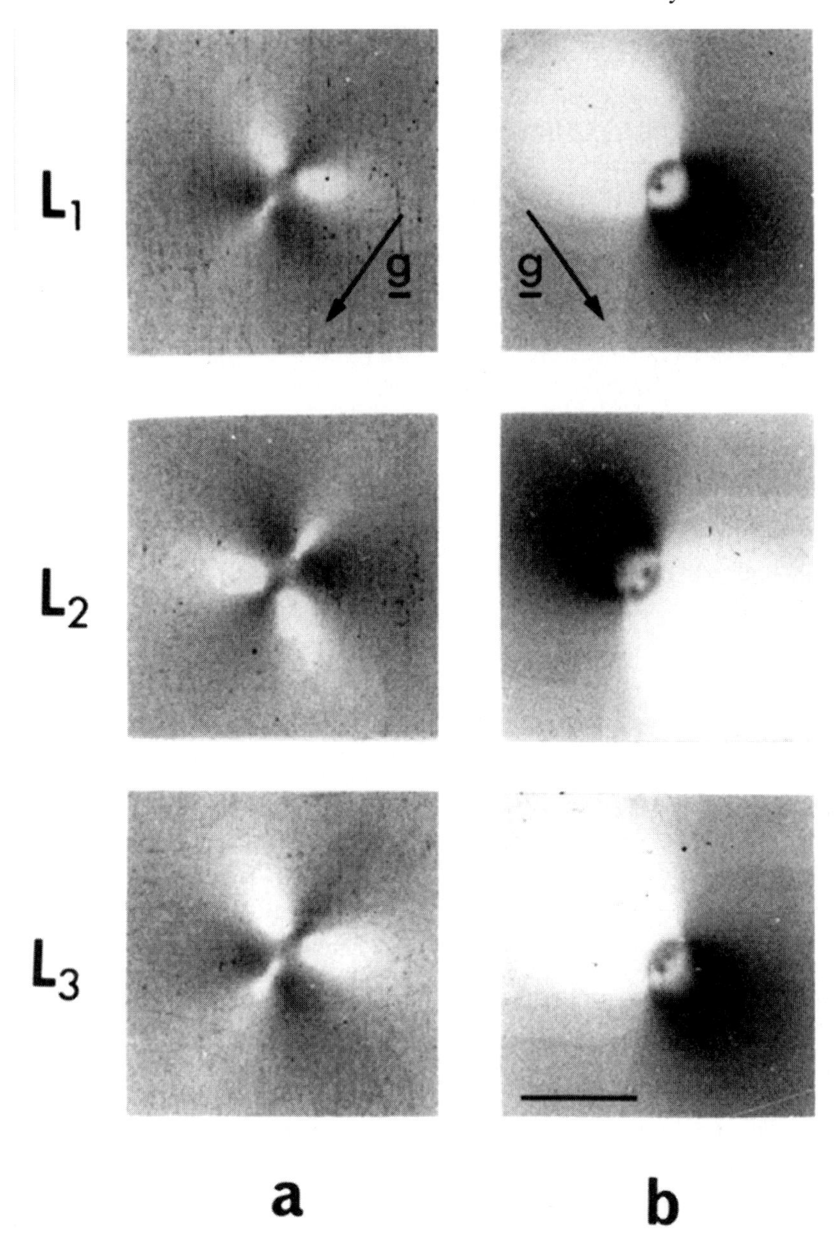

Figure 3.9. Dark-field simulated images of edge dislocation loop ($\boldsymbol{b} = \frac{1}{2}[11\bar{1}]$) in molybdenum, showing the calculated variation in contrast for loops located at the centres of depth layers L1, L2 and L3: (*a*) $\boldsymbol{g} = 2\bar{1}1$, $|\boldsymbol{g} \cdot \boldsymbol{b}| = 0$, type 2 contrast; (*b*) $\boldsymbol{g} = 21\bar{1}$, $|\boldsymbol{g} \cdot \boldsymbol{b}| = 2$, type 4 contrast. $z = [011]$, $s_g = 0$, $t = 5.25\xi_{110}$. Scale mark 10 nm. From Eyre *et al* (1977b).

3.1.3.6 Loops in transition zones

It will normally be the case that a fraction of the loops will lie in transition zones between depth layers. In early work these transition zones were called 'dead layers' with the implication that loops located at these depths would be invisible or show black-spot contrast with no black–white structure. Eyre *et al* (1977b) have shown that this is not necessarily the case. Calculations for edge-on and inclined dislocation loops showing type 3 contrast (columns (*a*) and (*c*)) and edge-on loops showing types 2 and 4 contrast (columns (*b*) and (*d*)) are shown in figure 3.10. In each column, the top and bottom images correspond to loops located at the centres of layers L1 and L2 respectively, while the other three images are for loops located close to or within the transition zone between L1 and L2. The calculations show that the loops in the transition zone are likely to be visible, and they often show rather complex structure as the black and white lobes interchange. There is sometimes room for confusion. For example, in the first column (*a*) the image of the edge-on loop located close to the centre of the transition zone shows contrast similar to type 4 whereas the $\boldsymbol{g} \cdot \boldsymbol{b}$ value would indicate that the contrast should be type 3. Such ambiguities can be removed in practice by taking micrographs in several different diffraction vectors or at different operating voltages. Since the extinction distance ξ_g varies with $\boldsymbol{g}$ and electron energy, the depth positions of the transition layers will change.

3.1.3.7 Effects of elastic anisotropy

Ohr (1979) has investigated the effects of elastic anisotropy in copper, nickel and niobium, and his main result is shown in figure 3.11. Simulations were made of the images of edge loops with Burgers vector $\boldsymbol{b}$ parallel to $[1\bar{1}1]$ imaged in a diffraction vector $\boldsymbol{g} = 002$ with foil normal $\boldsymbol{n} = [110]$. For a Frank loop in a hypothetical *isotropic* material, the computed image showed a black–white contrast with $\boldsymbol{l}$ rotated about 15° away from $\boldsymbol{b}$ in a direction towards $\boldsymbol{g}$. Calculations for Frank loops in copper and nickel were made under identical diffraction conditions using the displacement fields of anisotropic elasticity. In both cases the effect of anisotropy was to 'stretch' the black–white contrast parallel to the Burgers vector $\boldsymbol{b}$. The black–white vector $\boldsymbol{l}$ now aligned closely with the direction of the Burgers vector. The degree of elongation was higher in the more anisotropic copper.

These findings are in agreement with experimental observations. 'Stretched' black–white images with $\boldsymbol{l}$ parallel to $\boldsymbol{b}$ are indeed seen in copper, see figure 2.2. The close correspondence between $\boldsymbol{l}$ and $\boldsymbol{b}$ may explain why simple $\boldsymbol{l}$-vector analysis in copper and nickel has been so successful. Similar calculations for perfect edge loops in the less-anisotropic niobium showed that the black–white contrast direction $\boldsymbol{l}$ in this case lay approximately midway between $\boldsymbol{b}$ and $\boldsymbol{g}$. The contrast for the loops in copper and nickel is stretched in the $\langle 111 \rangle$ direction because this direction is elastically soft in these materials. In niobium, however,

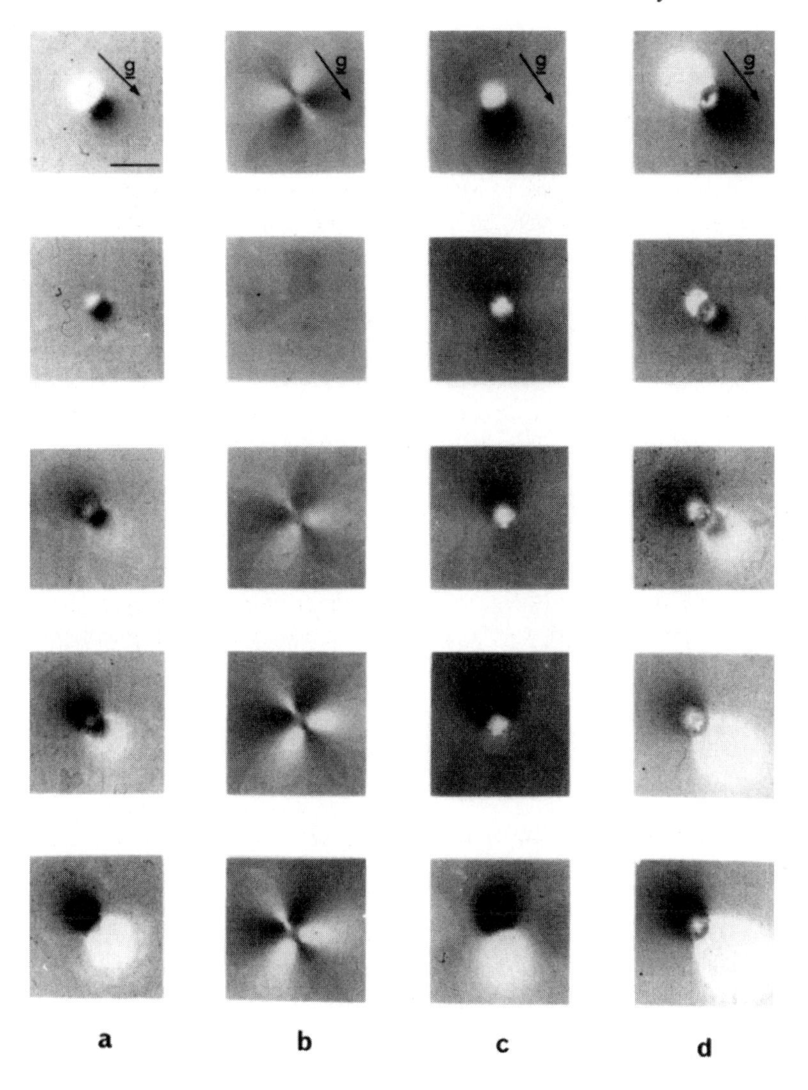

Figure 3.10. Dark-field simulated images ($s_g = 0$) for 'edge-on' ((*a*), (*b*) and (*d*)) and 'inclined' ((*c*)) edge dislocation loops in molybdenum, showing the calculated variations in contrast for loops located within the transition zone (middle three rows) between L1 (top row) and L2 (bottom row). (*a*) $z = [001]$, $g = 110$, $b = \frac{1}{2}[110]$, $|g \cdot b| = 1$ indicating type 3 contrast. (*b*) $z = [011]$, $g = 21\bar{1}$, $b = \frac{1}{2}[1\bar{1}1]$, $|g \cdot b| = 0$ indicating type 2 contrast. (*c*) $z = [011]$, $g = 21\bar{1}$, $b = \frac{1}{2}[111]$, $|g \cdot b| = 1$ indicating type 3 contrast. (*d*) $z = [011]$, $g = 21\bar{1}$, $b = \frac{1}{2}[11\bar{1}]$, $|g \cdot b| = 2$ indicating type 4 contrast. Foil thickness $t = 5.25\xi_{110}$ in all cases. In (*a*), depth of loop in foil from top to bottom $d = 5.1$, 5.0, 4.95, 4.9 and $4.75\xi_{110}$. In (*c*), (*b*) and (*d*), $d = 4.98$, 4.81, 4.72, 4.63 and $4.36\xi_{110}$. Scale mark 10 nm. From Eyre *et al* (1977b).

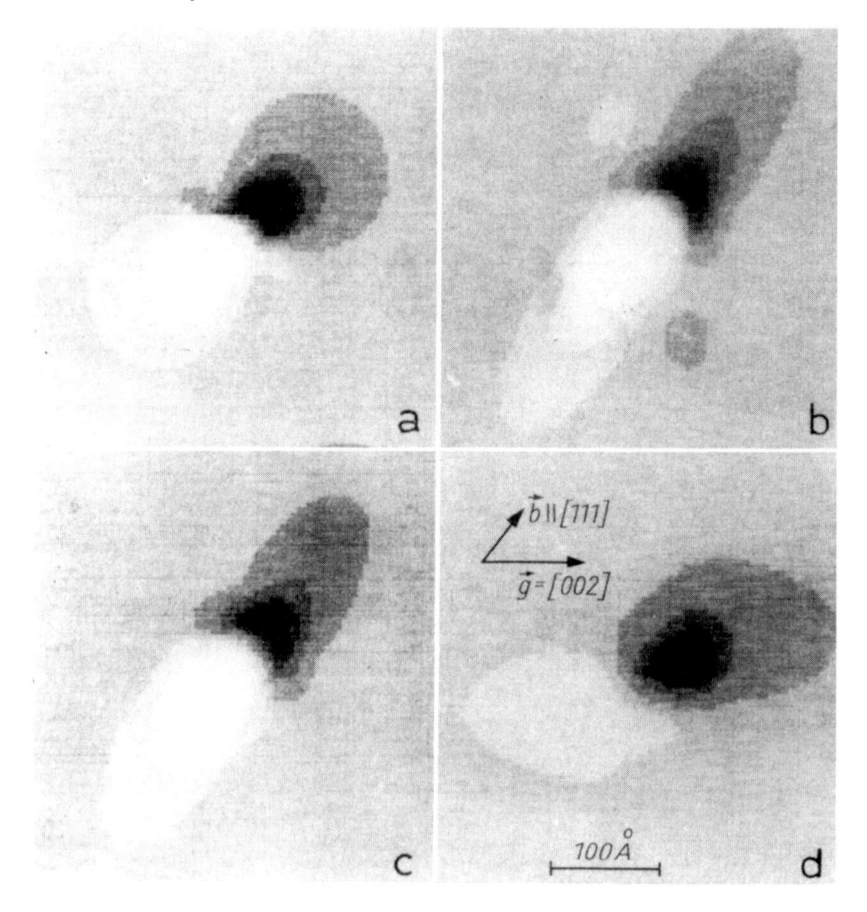

Figure 3.11. Simulated images of a pure edge $[1\bar{1}1]$ dislocation loop imaged in $\boldsymbol{g} = 002$ with $s_g = 0$ in (*a*) an isotropic fcc solid, (*b*) copper, (*c*) nickel and (*d*) niobium. In copper and nickel the black–white contrast is considerably elongated, and $\boldsymbol{l}$ is aligned parallel to $\boldsymbol{b}$. In niobium the black–white vector $\boldsymbol{l}$ lies midway between $\boldsymbol{g}$ and $\boldsymbol{b}$. From Ohr (1979).

this direction is elastically hard, and thus the black–white vector is rotated further from the $[1\bar{1}1]$ Burgers vector than for an elastically isotropic material.

3.1.3.8 Effects of shear components

Further complications arise if non-edge-type loops are present as is likely to be the case in bcc and hcp materials (see section 3.1.4). Holmes *et al* (1979) have made simulations of non-edge loops in molybdenum. They obtained the displacement field of a finite circular loop in an isotropic medium by adding, in appropriate proportions, the displacement field of a pure edge loop (Bullough and Newman

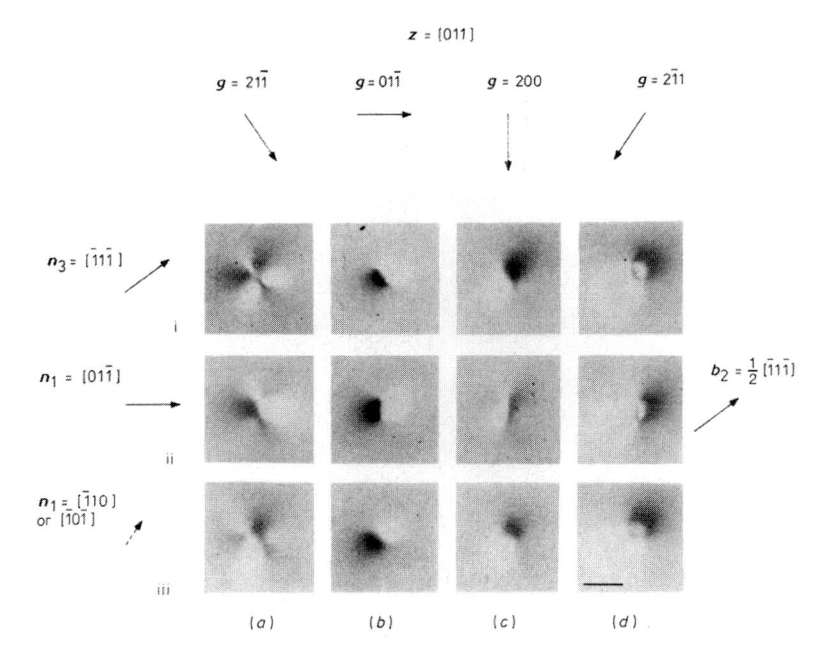

Figure 3.12. Dark-field simulated images of a loop with $\boldsymbol{b} = \frac{1}{2}[\bar{1}1\bar{1}]$, for $z = [011]$ and $s_g = 0$. Here the loop normals $\boldsymbol{n}$ are designated $\boldsymbol{n}_3$ if the loop is of edge type, or $\boldsymbol{n}_1$ if the loop habit-plane is the original nucleation plane. In row (i) the loop is pure edge with normal $\boldsymbol{n}_3 = [\bar{1}1\bar{1}]$. In row (ii) the loop is non-edge with normal $\boldsymbol{n}_1 = [01\bar{1}]$. In row (iii) the loop is non-edge with normal $\boldsymbol{n}_1 = [\bar{1}10]$ or $[\bar{1}0\bar{1}]$. Images are shown for the four diffraction vectors: (*a*) $\boldsymbol{g} = 21\bar{1}$, (*b*) $\boldsymbol{g} = 01\bar{1}$, (*c*) $\boldsymbol{g} = 200$ and (*d*) $\boldsymbol{g} = 2\bar{1}1$. Foil thickness $t = 5.25$, loop depth $d = 5.10$, loop radius 0.1, in units of ξ_{110}. Scale mark 10 nm. From English *et al* (1980).

1960) and a pure shear loop (Ohr 1972). They came to the following main conclusions on the effect of shear components of $\boldsymbol{b}$, illustrated in figures 3.12 and 3.13 (English *et al* 1980) which show simulations of loops with shear components in molybdenum with $\boldsymbol{b} = \frac{1}{2}[\bar{1}1\bar{1}]$ and with $\boldsymbol{b} = \frac{1}{2}[010]$ respectively. The top row of each set correspond to edge loops and show contrast of types 1–4 as previously described.

- *Type 1 images.* Images corresponding to $\boldsymbol{g} \cdot \boldsymbol{b} = 0$ which exhibit type 1 contrast in the pure edge configuration can show comparatively strong black–white contrast in non-edge configurations, with the development of a second white lobe (see figure 3.13(*c*), rows (ii) and (iii)).
- *Type 2 images.* $\boldsymbol{g} \cdot \boldsymbol{b} = 0$ images for $\boldsymbol{b}$ normal to z may be significantly distorted from the symmetrical type 2 butterfly images. There is a marked strengthening of one of the white lobes (figure 3.12(*a*), rows (ii) and (iii)).

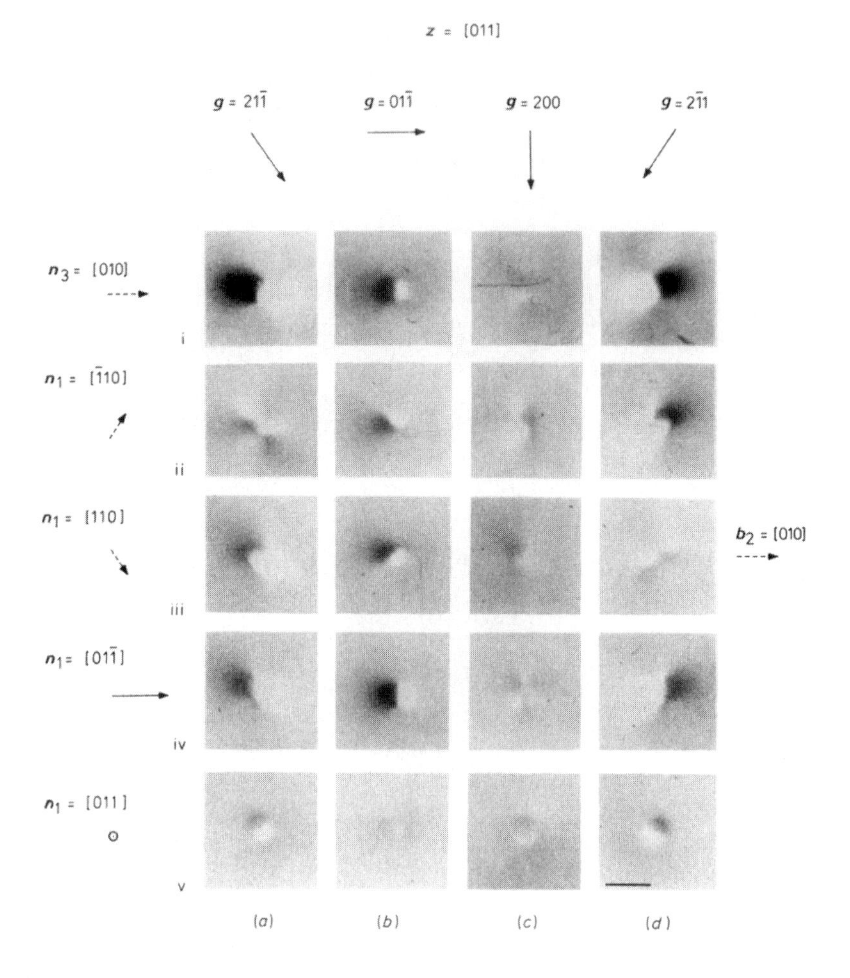

Figure 3.13. Dark-field simulated images of a loop with $\boldsymbol{b} = [010]$, for $z = [011]$ and $s_g = 0$. In row (i) the loop is pure edge with normal $\boldsymbol{n}_3 = [010]$. In row (ii) the loop is non-edge with normal $\boldsymbol{n}_1 = [\bar{1}10]$. In row (iii) the loop is non-edge with normal $\boldsymbol{n}_1 = [110]$. In row (iv) the loop is non-edge with normal $\boldsymbol{n}_1 = [01\bar{1}]$. In row (v) the loop is non-edge with normal $\boldsymbol{n}_1 = [011]$. Images are shown for the four diffraction vectors: (*a*) $\boldsymbol{g} = 21\bar{1}$, (*b*) $\boldsymbol{g} = 01\bar{1}$, (*c*) $\boldsymbol{g} = 200$ and (*d*) $\boldsymbol{g} = 2\bar{1}1$. Foil thickness $t = 5.25$, loop depth $d = 5.10$, loop radius 0.1, in units of ξ_{110}. Scale mark 10 nm. From English *et al* (1980).

This strong lobe tends to lie along $\boldsymbol{n}$ or its projection. This image feature allows the habit-planes of loops with $\boldsymbol{n}$ rotated out of the $\boldsymbol{b}$–z plane to be deduced. This is discussed in section 3.1.4.2.

- *Type 3 images* retain their basic form, with some skewing of the black–white lobes, and the black–white interface rotates to stay approximately normal to $\boldsymbol{n}$, see figure 3.12, columns (*b*) and (*c*). The black–white interface may develop a zig-zag or wavy structure for some combinations of $\boldsymbol{g}$, $\boldsymbol{b}$ and $\boldsymbol{n}$ (figure 3.13, (*a*) and (*d*), rows (ii) and (iii)).

- For some combinations of $\boldsymbol{g}$, $\boldsymbol{b}$, $\boldsymbol{n}$ and z it may be difficult to distinguish modified type 1 and 2 ($\boldsymbol{g} \cdot \boldsymbol{b} = 0$) images from modified type 3 ($0 < |\boldsymbol{g} \cdot \boldsymbol{b}| \leq 1$) images.
- *Type 4 images* retain their basic form and also retain their characteristic interface structure. There is some skewing of the black–white lobes in non-edge configurations, see figure 3.12(*d*). In a practical experiment, the recognition of these largely unchanged type 4 images allows relatively easy determinations of $\boldsymbol{b}$.
- If $\boldsymbol{n}$ and $\boldsymbol{b}$ lie close to the image plane, the normal $\boldsymbol{m}$ to the black–white interface in type 3 and 4 images provides a good estimate of $\boldsymbol{n}$. If $\boldsymbol{n}$ and $\boldsymbol{b}$ lie in the image plane, $\boldsymbol{m}$ is closely parallel to $\boldsymbol{n}$ (figure 3.12, columns (*b*), (*c*) and (*d*)).

Examples of the identification and analysis of loops with shear components are discussed further in section 3.1.4.

3.1.3.9 *Effects of surface relaxation*

In elastic strain field calculations, complete relaxation of the surface is possible for an infinitesimal loop and this has been done in some simulations (e.g. McIntyre 1967). Within the finite-loop model, it is also possible to relax the near-surface displacement field in order to make the surface stress-free. This is difficult to do exactly but the condition may be partially fulfilled for a loop parallel to the surface by taking the sum of the displacement fields of the loop and its mirror image in the foil surface. This was done in some early simulations (see, e.g., Rühle *et al* 1965). These authors were more concerned with effects of the surface on the layer structure (see p. 30) and did not attempt to examine systematically surface effects on loop contrast. Ohr (1977) has reported a more sophisticated technique for handling surface relaxation for finite loops, and has concluded that the image contrast of small loops lying very close to a stress-free surface is indeed sensitive to the presence of the surface. The surface was found to affect the size and detailed shape of the black–white contrast figure.

In practice, the boundary conditions at the surface are not well established. For example, a surface oxide layer may be present which constrains the underlying material. Provided the distance of the loop from the surface exceeds its diameter, neglect of surface effects does not appear to be too serious a problem.

Surface relaxation does, however, have an effect on the extent of the first black–white depth layer L1. The reason for this is clear. Surface relaxation tends to make $\beta = \mathrm{d}(\boldsymbol{g} \cdot \boldsymbol{R})/\mathrm{d}z$ symmetrical, and so expands the first layer. For a finite loop, the first layer is predicted to be of thickness $0.3\xi_g$ when surface relaxation is included, compared with $0.25\xi_g$ when it is not. It is not always clear if surface relaxation occurs in practice. This introduces some uncertainty about the extent of the first layer which has some implications when the black–white stereo technique is used to determine the nature of the loop (see section 4.2).

3.1.3.10 *Effects of increasing s_g*

As the deviation parameter s_g increases, the visibility of black–white contrast decreases. This effect can be seen experimentally in figure 2.4. In addition, the effective extinction distance ξ_g^{eff} decreases, so that the layer structure becomes compressed.

For large values of s_g weak-beam conditions are attained. This will be discussed in section 3.2.

3.1.4 Examples of image matching to determine Burgers vectors and habit-planes

3.1.4.1 *Burgers vector determination*

The application of the theoretical predictions described in section 3.1.3 to the analysis of a mixed population of $\frac{1}{2}\langle 111\rangle$ and $\langle 100\rangle$ dislocation loops in ion-irradiated molybdenum has been described by English *et al* (1980). Figure 3.14 is taken from this work. The centre row of the figure shows experimental images of four loops, labelled A–D, obtained using two $\langle 211\rangle$ diffraction vectors. The top and bottom rows show simulated images of two types of loops in both edge and non-edge configurations. The top row shows images for loops with $\boldsymbol{b} = \frac{1}{2}[\bar{1}1\bar{1}]$. For columns (*a*) and (*b*) the loops are in edge orientation with $\boldsymbol{n}_3 = [\bar{1}1\bar{1}]$[4] whilst in columns (*c*) and (*d*) they are non-edge with $\boldsymbol{n}_1 = [\bar{1}10]$. The bottom row shows corresponding images for $\boldsymbol{b} = [100]$. In columns (*a*) and (*b*), $\boldsymbol{n}_3 = [100]$ and in columns (*c*) and (*d*) $\boldsymbol{n}_1 = [\bar{1}10]$.

The symmetries seen in the experimental images immediately allow the Burgers vectors of the loops to be deduced even without detailed image matching. Since all four loops show type 4 contrast in $\boldsymbol{g} = 21\bar{1}$, the possible Burgers vectors are either $\boldsymbol{b} = \frac{1}{2}[\bar{1}1\bar{1}]$ or $\boldsymbol{b} = [100]$, or the faulted types $\boldsymbol{b} = \frac{1}{2}[\bar{1}10]$ or $\frac{1}{2}[101]$. The image in $\boldsymbol{g} = 21\bar{1}$ allows these possibilities to be differentiated. In this reflection, defects A and B retain type 4 contrast corresponding to $\boldsymbol{g}\cdot\boldsymbol{b} > 1$ and so have $\boldsymbol{b} = [100]$. Defects C shows type 2 'butterfly' contrast so that $\boldsymbol{g}\cdot\boldsymbol{b} = 0$, implying $\boldsymbol{b} = \frac{1}{2}[\bar{1}1\bar{1}]$. Defect D shows distorted type 2 contrast, suggesting its Burgers vector is also $\boldsymbol{b} = \frac{1}{2}[\bar{1}1\bar{1}]$. This assignation is confirmed by detailed image matching, where it is seen that a good match is obtained for a non-edge loop with $\boldsymbol{b} = \frac{1}{2}[\bar{1}1\bar{1}]$ lying on $(\bar{1}10)$.

[4] Loop normals $\boldsymbol{n}$ are designated $\boldsymbol{n}_3$ if the loop is of edge type, or $\boldsymbol{n}_1$ if the loop habit-plane is the original nucleation plane, see figure 3.12.

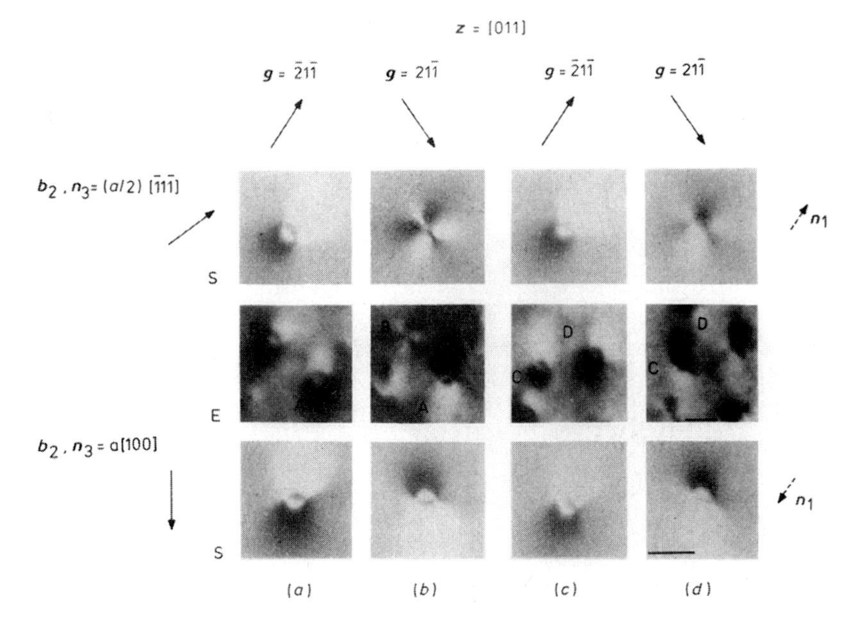

Figure 3.14. Comparison of experimental (E) and simulated (S) dark-field images of edge and non-edge perfect loops in irradiated molybdenum. In the top row, $\boldsymbol{b} = \frac{1}{2}[\bar{1}1\bar{1}]$, and in (*a*) and (*b*) $\boldsymbol{n}_3 = [\bar{1}1\bar{1}]$, and in (*c*) and (*d*) $\boldsymbol{n}_1 = [\bar{1}10]$. In the bottom row, $\boldsymbol{b} = [100]$, and in (*a*) and (*b*) $\boldsymbol{n}_3 = [100]$, and in (*c*) and (*d*) $\boldsymbol{n}_1 = [\bar{1}10]$. The experimental micrographs in the centre row were obtained in an [011] molybdenum foil irradiated with 180 keV Sb_3^+ ions to a dose of 10^{16} ion m^{-2}. Images are shown for the two diffraction vectors: (*a*) and (*c*) $\boldsymbol{g} = \bar{2}1\bar{1}$; (*b*) and (*d*) $\boldsymbol{g} = 21\bar{1}$. Scale marker 10 nm. From English *et al* (1980).

This example shows that relatively simple experiments are capable of unambiguously distinguishing loops of different Burgers vectors, even when non-edge loops are present.

Mixed populations of $\boldsymbol{b} = \frac{1}{2}\langle 111\rangle$ and $\langle 100\rangle$ dislocation loops have also been found in ion- and neutron-irradiated iron and ferritic steels (e.g. Jenkins and English 1978, Robertson *et al* 1982, Horton *et al* 1982, Little 1972). This has important implications for understanding void swelling in iron and ferritic steels. One mechanism leading to void swelling is the so-called dislocation bias: the elastic interaction between dislocations and point-defects is greater for interstitials than vacancies, leading to preferential absorption of interstitials. Under such conditions vacancies can aggregate into neutral sinks such as voids. Dislocation loops also act as biased sinks. However, since the elastic interaction with interstitials depends on $|\boldsymbol{b}|$, when both $\langle 100\rangle$ and $\frac{1}{2}\langle 111\rangle$ loops are present, the former type tend to be the dominant sinks for mobile interstitials. Under such circumstances, the $\frac{1}{2}\langle 111\rangle$ loops act as relatively neutral sinks. This tends to suppress the nucleation of voids and so reduces swelling. Details are given by Little *et al* (1980).

3.1.4.2 Habit-plane determination

The previous example clearly shows that black–white images give information on the loop habit-planes. In favourable cases, the loop normal $\boldsymbol{n}$ may be estimated from the direction of the normal, $\boldsymbol{m}$, to the black–white interface for type 3 and 4 images. $\boldsymbol{n}$ is parallel to $\boldsymbol{m}$ if $\boldsymbol{b}$ and $\boldsymbol{n}$ are both in the image plane. Even if $\boldsymbol{b}$ and $\boldsymbol{n}$ are not both in the image plane, $\boldsymbol{m}$ may still give a reasonable estimate of $\boldsymbol{n}$ although the rather subtle contrast changes with varying $\boldsymbol{n}$ may make an unambiguous determination difficult.

If we wish to analyse loops in materials such as molybdenum or iron some crystallographic considerations are helpful. Perfect loops in bcc crystals are generally believed to nucleate as faulted loops with $\boldsymbol{b} = \frac{1}{2}\langle 110\rangle$ on $\{110\}$ planes (Eyre and Bullough 1965). Since the stacking-fault energy is high, such loops are subsequently expected to unfault to perfect loops with $\boldsymbol{b} = \frac{1}{2}\langle 111\rangle$ or $\boldsymbol{b} = \langle 100\rangle$. These loops would also initially lie on $\{110\}$ planes, but would be expected to reduce their line energy by rotating on their glide cylinders towards edge orientation. This picture is clearly in accord with the observations about figure 3.14. We might also expect loops to be present with $\boldsymbol{n}$ lying at some angle θ intermediate between the nucleation plane normal and $\boldsymbol{b}$, forming a 'family' of loops with the same Burgers vector $\boldsymbol{b}$ and original nucleation plane but varying θ.

The image characteristics for loops with shear components described in section 3.1.3 provide a general basis for determining the habit-planes of non-edge loops. The recipe recommended by English *et al* (1980) is as follows. First, the Burgers vectors of the loops are determined and loops with a given $\boldsymbol{b}$ chosen for habit-plane analysis. Then the relative strengths of the white lobes in distorted type 2, $\boldsymbol{g} \cdot \boldsymbol{b} = 0$ images are used to distinguish between the different possible families of loops with this Burgers vector, and hence to determine the particular nucleation plane of a given family. Finally, image matching is carried out for loops in this family. Matching is carried out by comparing experimental images with simulated images showing modified type 3 and type 4 contrast, with $\boldsymbol{n}$ lying at varying angles θ between the nucleation plane normal and the normal to the pure edge plane. It is advantageous in practice if the image plane contains $\boldsymbol{b}$ (i.e. $\boldsymbol{b} \cdot \boldsymbol{z} = 0$). The procedure is illustrated in figure 3.15 for three loops with $\boldsymbol{b} = \frac{1}{2}[11\bar{1}]$. A good match is found in each column. The loop in the centre column has its loop normal $\boldsymbol{n}_2$ halfway between $\boldsymbol{n}_1 = [01\bar{1}]$ and $\boldsymbol{n}_3 = [11\bar{1}]$.

A further example of loop habit-plane determination by this method is shown in figure 3.16, taken from a study of heavy-ion damage in ruthenium by Phythian *et al* (1987). The bottom row of the figure shows experimental images of three loops with $\boldsymbol{b} = \frac{1}{3}[2\bar{1}\bar{1}0]$ taken in diffraction vectors such that $\boldsymbol{g} \cdot \boldsymbol{b} = 0$, 1 and 2, respectively. Simulated images for different loop normals $\boldsymbol{n}$, with the angle θ between $\boldsymbol{n}$ and $\boldsymbol{b}$ increasing from zero to 30°, are shown in the top four rows.

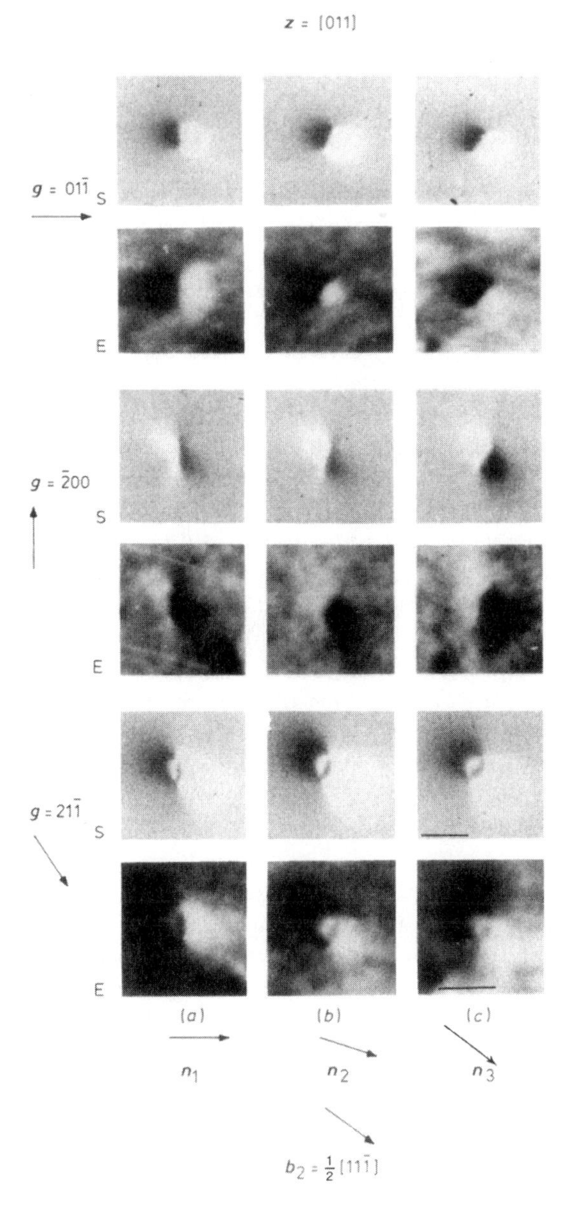

Figure 3.15. Comparison of experimental dark-field images (E) with simulated images (S) of three loops with $\boldsymbol{b} = \frac{1}{2}[11\bar{1}]$ for $z = [011]$ and $s_g = 0$, for the three diffraction vectors shown. In column (*a*) $\boldsymbol{n}_1 = [01\bar{1}]$; in column (*b*) $\boldsymbol{n}_2$ is halfway between $[01\bar{1}]$ and $[11\bar{1}]$ and in column (*c*) $\boldsymbol{n}_3 = [11\bar{1}]$. The experimental images were obtained in an [011] molybdenum foil irradiated with 60 keV W^+ ions to a dose of 5×10^{16} ion m^{-2} at 400 °C. Scale marker 10 nm. From English *et al* (1980).

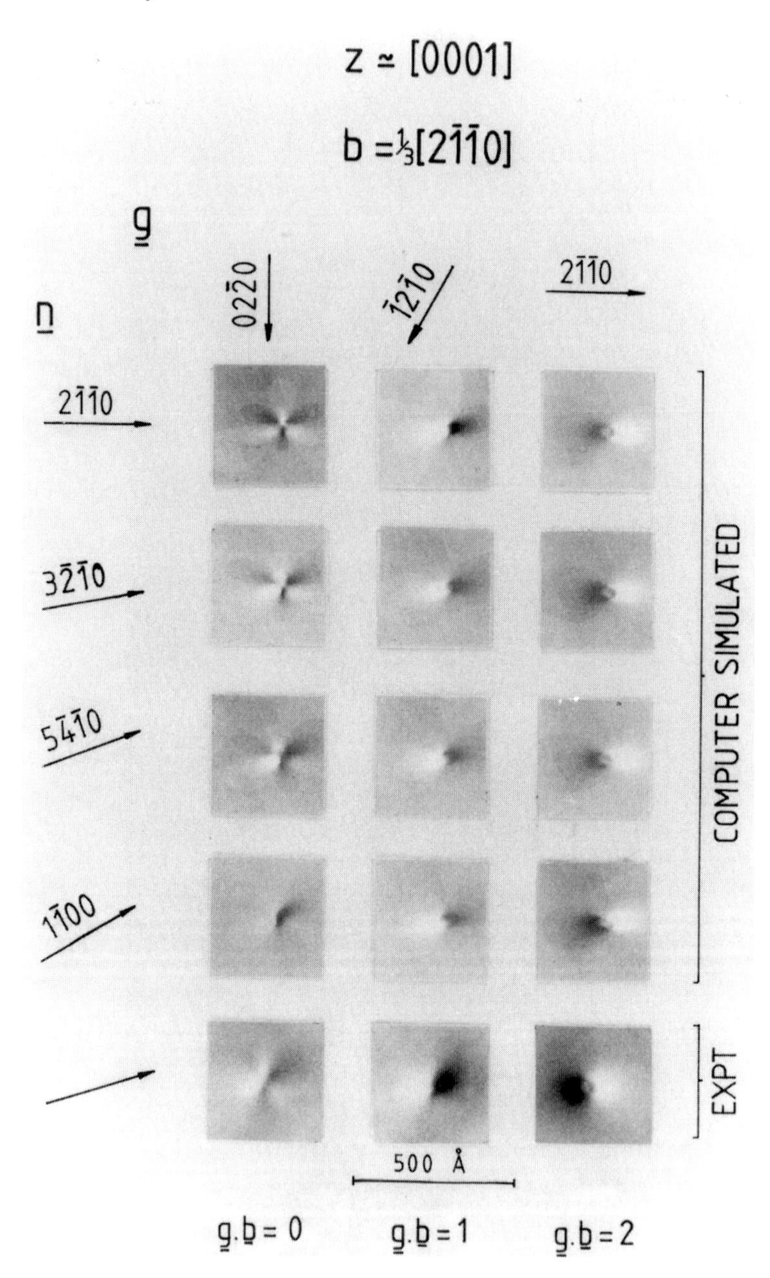

Figure 3.16. Comparisons of experimental and computer-simulated images showing the determination of the habit-plane of a loop in ion-irradiated ruthenium. The four sets of computed images at the top are for different loop normals $\boldsymbol{n}$ for the three reflections shown. The best match is for $\boldsymbol{n}$ close to $5\bar{4}\bar{1}0$. From Phythian *et al* (1987).

It can be seen that the images in this hcp material have similar characteristics to images in bcc materials, and can be classified as distorted type 2, 3 or 4 for the three columns, respectively. The best match is for the centre two rows of simulated images, corresponding to $\theta \approx 20^\circ$. From this and similar experiments, it may be concluded that the plane normal $\boldsymbol{n}$ may be determined to within 5–10°, the uncertainty being due to matching the fairly subtle distortions in contrast which arise as $\boldsymbol{n}$ is varied.

In this and other hcp metals with a less than ideal c/a ratio (Ti, Co, Re and Mg) it would appear that vacancy loops produced by cascade collapse generally nucleate as faulted loops on prism planes. Such loops may unfault to perfect loops because of the high stacking-fault energy, by a mechanism analagous to that found in bcc metals (Föll and Wilkens 1975):

$$\frac{1}{2}[10\bar{1}0] + \frac{1}{6}[1\bar{2}10] = \frac{1}{3}[2\bar{1}\bar{1}0].$$

These perfect loops may subsequently rotate on their glide cylinders towards the edge orientation in order to lessen their line length.

3.1.5 Black–white images of SFT

SFT have a weak symmetric long-range strain field and so would be expected to show black–white contrast similar to isotropic misfitting precipitates. This in fact is so: both experiments and contrast calculations confirm that small SFT in the layer structure do indeed show black–white contrast with $\boldsymbol{l}$ parallel to $\boldsymbol{g}$ (Saldin *et al* 1979). The simulations carried out by these authors employed a technique whereby the strain field of the SFT was constructed from a superposition of the elastic strain fields of the angular dislocations. A comparison between experimental and computed images is shown in figure 3.17. Good agreement is found between simulations and experiment. Black–white contrast is seen even for low-order reflections, and $\boldsymbol{l}$ lies close to $\boldsymbol{g}$.

It can be seen that the geometry of the SFT leads to characteristic shapes of the interface separating the black and white lobes. The contrast is often dominated by the stacking faults. In certain orientations and imaging conditions larger SFT may show a pronounced V-shape of the line dividing black from white in dynamical images, with the lines of the V lying along the projections of steeply inclined or edge-on $\{111\}$ planes, see, for example, figure 3.17, column I (a_2). However, many defects giving similar contrast to such complete tetrahedra can be shown by weak-beam microscopy to be in fact partially-dissociated Frank loops—an intermediate configuration between a Frank loop and an SFT (see section 3.2). In practice SFT can often be more easily recognized under kinematical imaging conditions from the triangular contrast which they show in many orientations, or square contrast in $\langle 100\rangle$ orientations (see, for example, figure 2.3).

Some recent simulations of the contrast of SFT under weak-beam diffraction conditions are mentioned briefly in section 3.3, figure 3.27.

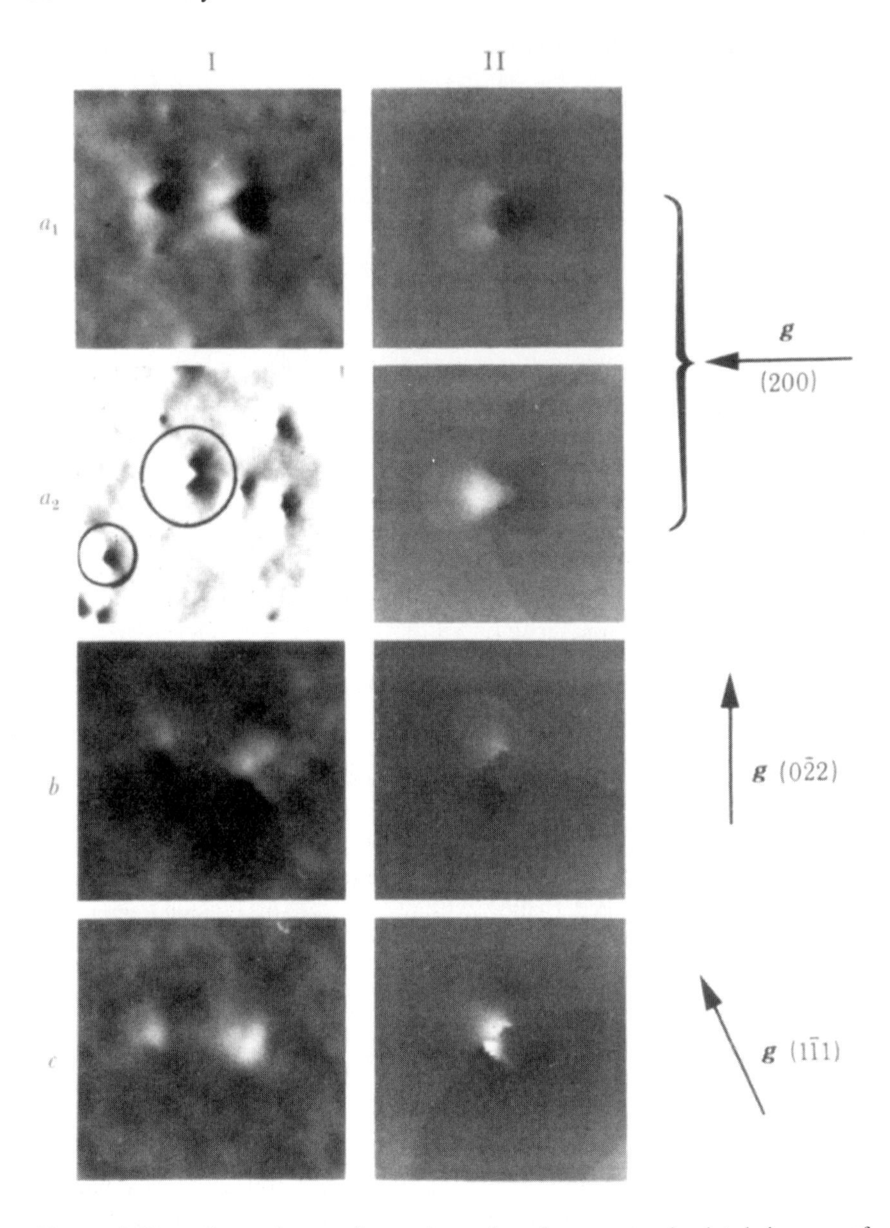

Figure 3.17. Comparisons of experimental and computer-simulated images of stacking-fault tetrahedra in copper. Experimental dark-field ($s_g = 0$) images are shown in column I, and computer simulated images in column II , for the diffraction vectors shown. The foil is oriented close to [011]. (a_1) and (a_2) show SFT in two different orientations, related by a rotation of 180° about [011]. The SFT in (*b*) and (*c*) are in the same orientation as (a_1). From Saldin *et al* (1979).

3.2 Weak-beam imaging

It was explained in chapter 2 that weak-beam imaging should always be carried out with the dimensionless deviation parameter $w = s_g \xi_g \geq 5$. In this case, the effective extinction distance is given by (Hirsch *et al* 1977)

$$\xi_g^{\text{eff}} = \frac{\xi_g}{\sqrt{1+w^2}} \approx \frac{1}{|s_g|}.$$

For the recommended value of $|s_g| = 2 \times 10^{-1}$ nm^{-1}, $\xi_g^{\text{eff}} = 5$ nm. In weak-beam imaging of small clusters, the size of the clusters relative to ξ_g^{eff} is important. The contrast of defects smaller than ξ_g^{eff} is likely to be more sensitive to parameters such as s_g, the depth of the defect in the foil and the foil thickness. For example, we have shown experimentally that the intensity of the weak-beam contrast of very small loops can fluctuate with small changes in $|s_g|$ and may sometimes be so weak that the defect is invisible. For larger defects such variations are likely to be less important. This is indeed found in practice. Weak-beam imaging has been very successful in studies of the geometry of dislocation loops and more complex clusters of size larger than about 5 nm. Some examples are given in the following section. Weak-beam imaging has also been used successfully in studies of defects smaller than 5 nm, but in these cases the variability of the contrast with the exact diffracting conditions has to be taken into account. This will be discussed in section 3.2.2.

3.2.1 Weak-beam analyses of clusters of size >5 nm

Weak-beam imaging has proved very successful in the analysis of point-defect clusters of size >5 nm. We describe here two examples which demonstrate the usefulness of the weak-beam technique for this type of investigation.

Example 1: Frank loops in silicon

The conditions under which weak-beam images can give useful information on the geometry, size and nature of small Frank loops has been investigated by Jenkins *et al* (1973). Suitable experimental conditions were deduced from computed weak-beam images of Frank loops under various selected loop and foil orientations and diffraction vectors. Experiments to check these deductions were carried out on small interstitial Frank loops in silicon formed by implantation with 80 keV P^+ ions, followed by annealing at 700 °C for 1 hr. A series of weak-beam micrographs taken under various diffraction conditions are shown in figure 3.18. The foil orientations are shown by the inset Thompson tetrahedra (see the appendix for details of Thompson tetrahedron notation). The main conclusions were as follows.

- Loops of size greater than about 8 nm show consistent, interpretable contrast under weak-beam conditions with $|s_g| \geq 2 \times 10^{-1}$ nm^{-1}. The calculations suggested that loops of size down to about 5 nm might be amenable to simple analysis, although in practice there are difficulties towards the lower end of the range 5–8 nm.

- The four Frank loop variants in a (111) foil could easily be distinguished by weak-beam imaging using the three $\boldsymbol{g} = \langle 220 \rangle$ diffraction vectors. In any one reflection, say $\boldsymbol{g} = \bar{2}20$, figure 3.18(*a*)–(*c*), two sets of loops lying on the inclined planes α and β with $\boldsymbol{g} \cdot \boldsymbol{b} \neq 0$, i.e. $\boldsymbol{b} = \frac{1}{3}[\bar{1}1\bar{1}]$ (marked A) and $\boldsymbol{b} = \frac{1}{3}[1\bar{1}\bar{1}]$ (marked B) appear in strong contrast. The loops lying on the third inclined plane γ with $\boldsymbol{b} = \frac{1}{3}[\bar{1}\bar{1}1]$ have $\boldsymbol{g} \cdot \boldsymbol{b} = 0$ and are not visible. However, this third set C can be imaged in another {220} reflection, say $\boldsymbol{g} = 0\bar{2}2$, figure 3.18(*d*).
- The fourth set of loops, those marked D lying in the foil plane δ with $\boldsymbol{b} = \frac{1}{3}[111]$ are of particular interest. For any of the {220} reflections at the [111] pole, $\boldsymbol{g} \cdot \boldsymbol{b} = 0$. However, those portions of the loops with line direction $\boldsymbol{u}$ not lying parallel to $\boldsymbol{g}$ have $\boldsymbol{g} \cdot \boldsymbol{b} \times \boldsymbol{u} \neq 0$ and are visible. Calculations show that these images define both the shapes and sizes of the loops. In this case the loops are hexagonal, with edges along the $\langle 110 \rangle$ directions.
- The loops on the inclined {111} planes show inside–outside contrast changes with change in the sign of $\boldsymbol{g}$ (figure 3.18(*b*)) or the sign of s_g (figure 3.18(*c*)). These contrast changes allow the interstitial nature of the loops to be established (see section 4.1).
- The faulted nature of the majority of the loops can be confirmed by imaging using a diffraction vector not in the foil plane, e.g. $\boldsymbol{g} = \pm\bar{1}11$ (figure 3.18(*e*) and (*f*)). The stacking faults enclosed by loops lying in the foil plane δ appear as solid white areas, which very clearly show the hexagonal shape of the loops when the contrast of the bounding Frank partial dislocation lies inside the stacking-fault contrast, as in figure 3.18(*e*). When the dislocation contrast lies outside the stacking-fault contrast, as in figure 3.18(*f*), the shapes of the loops are less clearly delineated. Loops on γ show stacking-fault fringes. Edge-on loops on α and β appear as straight lines. A few perfect loops (marked P) not showing stacking-fault contrast are immediately evident in figure 3.18(*e*) and (*f*).

Example 2: Partially dissociated Frank loops in copper

In example 1, the loop geometry is rather simple, and simple contrast criteria (in particular the $\boldsymbol{g} \cdot \boldsymbol{b} = 0$ and $\boldsymbol{g} \cdot \boldsymbol{b} \times \boldsymbol{u} = 0$ invisibility criteria) could be used successfully to predict the observed contrast. The same may be true even when the defect geometry is more complicated. A good example of this is in the weak-beam analysis of partially-dissociated Frank loops (Jenkins 1974). In this work weak-beam electron microscopy was used to develop a model of the geometry of the point-defect clusters in silver and copper produced by heavy-ion irradiation.

A weak-beam image of ion-damage clusters in silver is shown in figure 3.19. The (111) foil orientation is shown by an inset Thompson tetrahedron, and

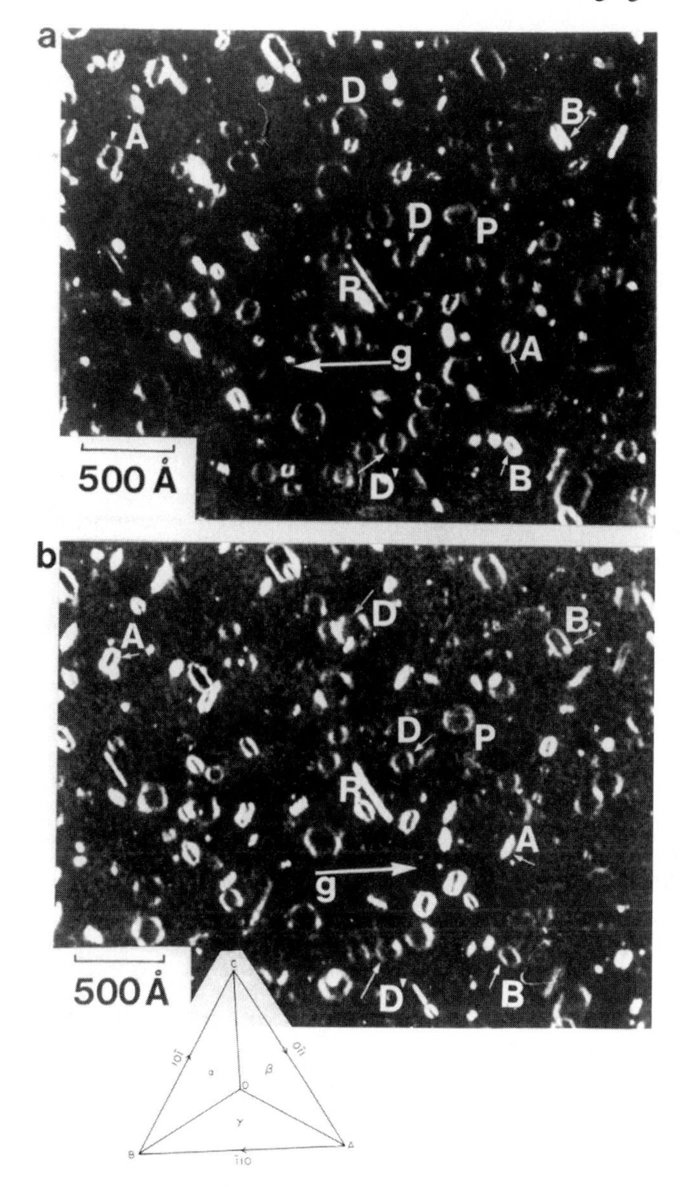

Figure 3.18. Weak-beam images of Frank loops in a (111) foil of silicon. The loops formed after implantation with 80 keV P^+ ions, followed by annealing at 700 °C for 1 hr: (*a*) $\boldsymbol{g} = \bar{2}20$, $s_g = 2 \times 10^{-1}$ nm^{-1}; (*b*) $\boldsymbol{g} = 2\bar{2}0$, $s_g = 2 \times 10^{-1}$ nm^{-1}; (*c*) $\boldsymbol{g} = 2\bar{2}0$, $s_g = -2 \times 10^{-1}$ nm^{-1}; (*d*) $\boldsymbol{g} = 0\bar{2}2$, $s_g = 2 \times 10^{-1}$ nm^{-1}; (*e*) $\boldsymbol{g} = \bar{1}11$, $s_g = -2 \times 10^{-1}$ nm^{-1}; (*f*) $\boldsymbol{g} = 1\bar{1}\bar{1}$, $s_g = -2 \times 10^{-1}$ nm^{-1}. Determination of the nature of these loops by the inside outside technique is discussed in section 4.1: they turn out to be interstitial. From Jenkins *et al* (1973).

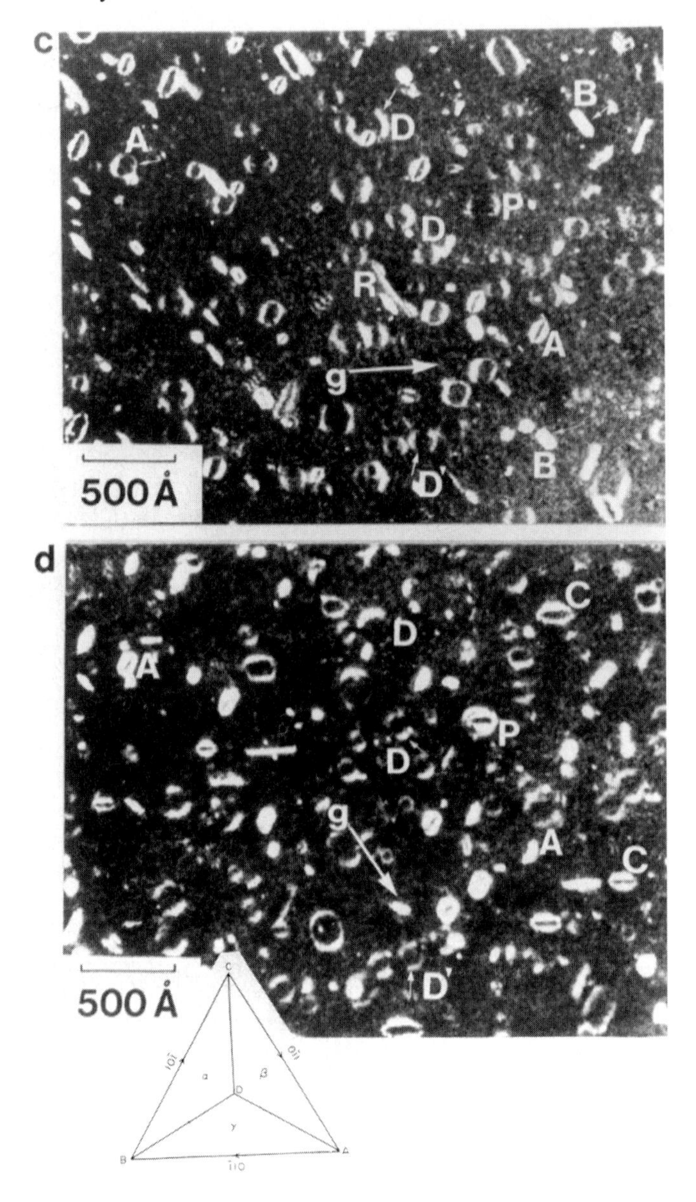

Figure 3.18. (Continued)

$g = \bar{2}20$. Clusters showing characteristic image features are labelled A, B, C and T. The way in which these characteristic features change when imaged in different $\{\bar{2}20\}$ diffraction vectors are shown in figures 3.20–3.22. In each case the weak-beam images have $s_g = 2 \times 10^{-1}$ nm^{-1}. The contrast changes seen in

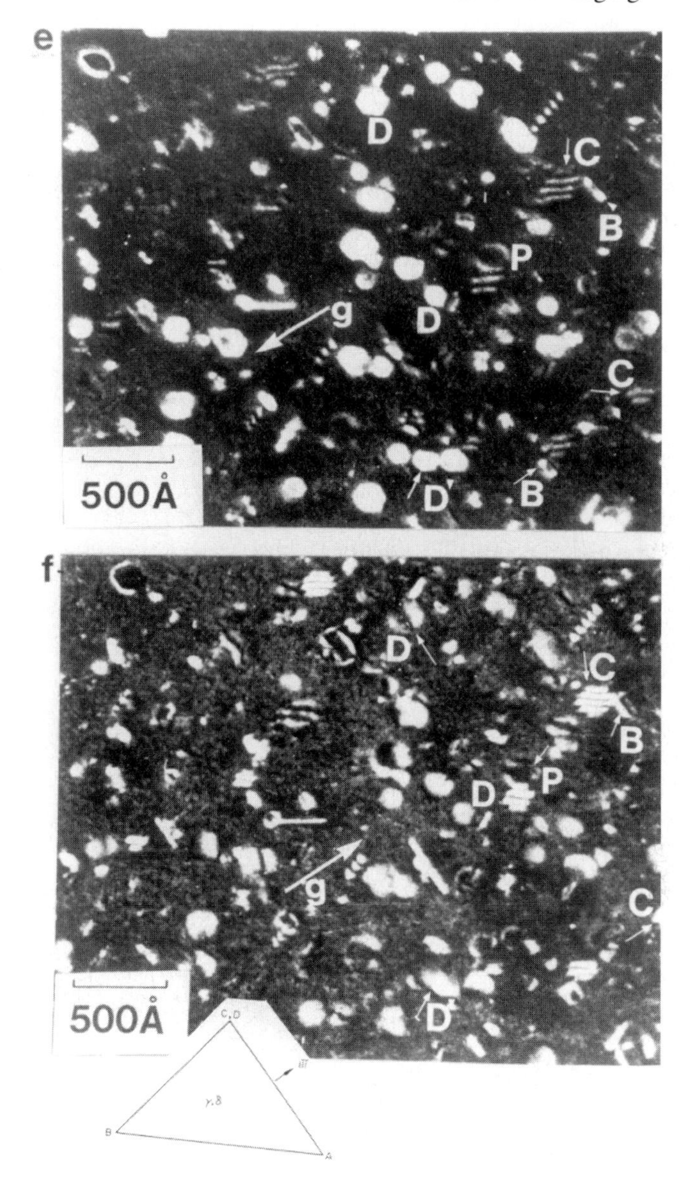

Figure 3.18. (Continued)

figures 3.20–3.22 are shown schematically in figure 3.23. The weak-beam images have the following features.

- Some of the clusters show the contrast expected from small SFT (e.g. the defects labelled T, and resembling arrowheads, in figure 3.19). The contrast

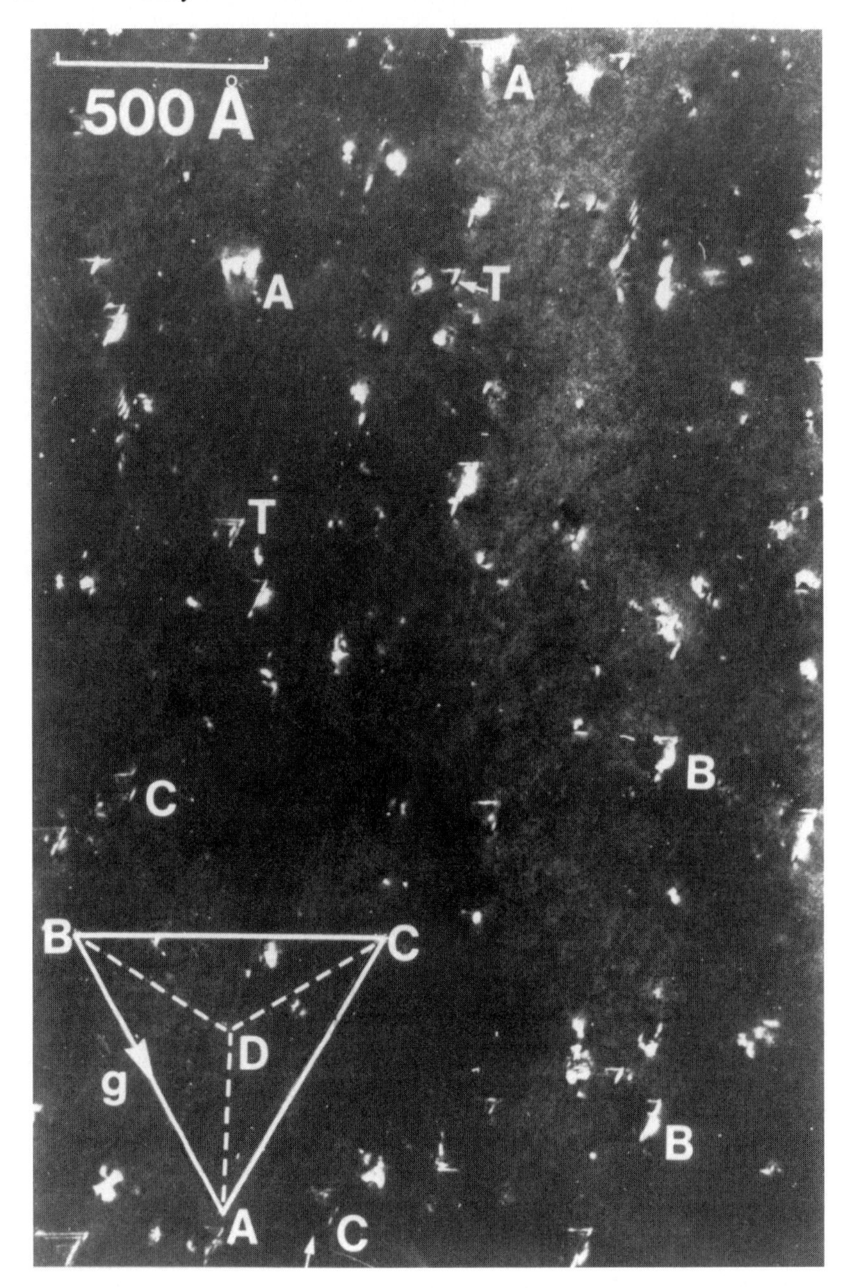

Figure 3.19. A weak-beam image ($\boldsymbol{g} = 2\bar{2}0$, $s_g = 2 \times 10^{-1}$ nm^{-2}) of vacancy clusters in silver produced by heavy-ion bombardment. Some characteristic contrast features are labelled A, B, C and T. Some contrast experiments showing how these features change with imaging conditions are shown in the following three figures. From Jenkins (1974).

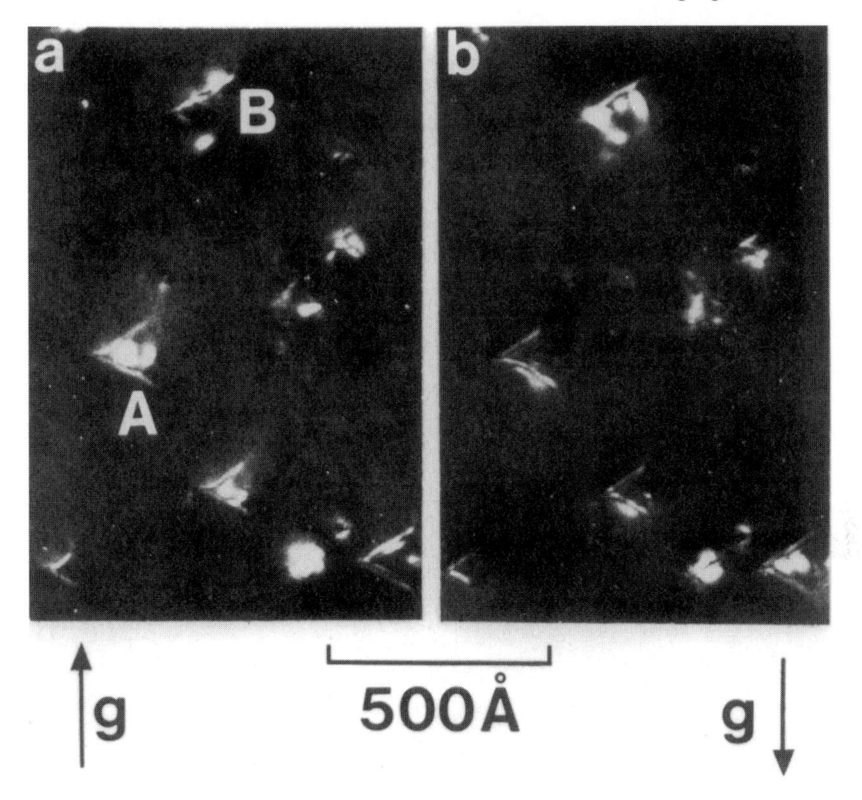

Figure 3.20. A contrast experiment on defects in heavy-ion irradiated silver—the effect of changing the sign of $\boldsymbol{g}$: (*a*) $\boldsymbol{g} = 2\bar{2}0$; (*b*) $\boldsymbol{g} = \bar{2}20$, both with $s_g = 2 \times 10^{-1}$ nm^{-1} and the foil oriented close to [111]. The feature labelled 'A', which shows (a)-type contrast, in figure 3.20(*a*) shows '(b)-type' contrast in figure 3.20(*b*), and *vice versa*. From Jenkins (1974).

arises predominantly from the inclined faces of the stacking fault. If more than one fringe is visible (see figure 5.4), the depth periodicity of the fringes is the effective extinction distance $\xi_g^{\text{eff}} = |s_g|^{-1} = 5$ nm. As $\boldsymbol{g}$ is changed, different faces of the SFT show contrast.

- Two different orientations of SFT are possible, with apexes pointing away from or towards the ion-entry surface. With few exceptions, only the first orientation is seen. Since the clusters lie within about 10 nm of the foil surface, this suggests that the surface plays an important role in determining the cluster configuration.
- Other clusters of size >8 nm show more complex contrast, which can be categorized into three different types—(a), (b) and (c)—shown diagrammatically in figure 3.23. In the micrographs of figures 3.19–3.22,

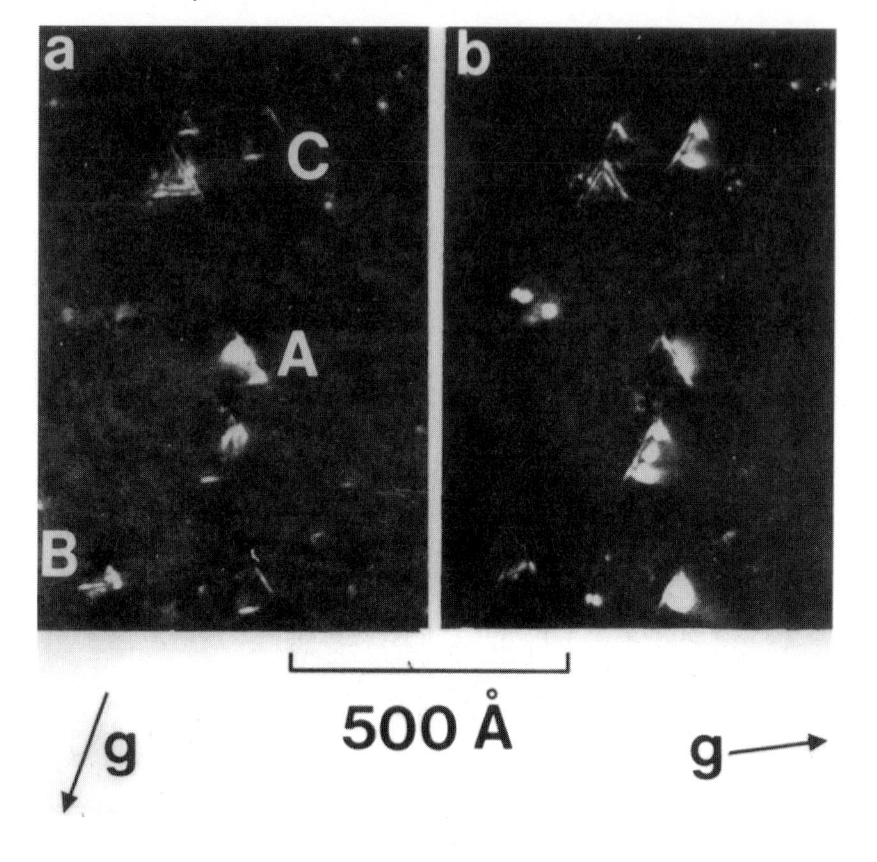

Figure 3.21. A contrast experiment on defects in heavy-ion irradiated silver—the effect of changing the diffraction vector: (*a*) $\boldsymbol{g} = 2\bar{2}0$; (*b*) $\boldsymbol{g} = \bar{2}02$, $s_g = 2 \times 10^{-1}$ nm^{-1} and the foil oriented close to [111]. Contrast changes are: (a)-type → (b)-type; (b)-type → (c)-type; (c)-type → (a)-type. From Jenkins (1974).

clusters showing the contrast of each of these types are labelled A, B and C. The contrast of individual clusters changes in consistent ways with changes in $\boldsymbol{g}$. For example type (a) contrast changes to type (b) contrast with a change in the sign of $\boldsymbol{g}$ (figure 3.20), while (c), (a) and (b) type contrast interchange with a change in the direction of $\boldsymbol{g}$ (figure 3.21).

- The contrast shown by a particular cluster is relatively insensitive to the exact value of $|s_g|$.
- The black–white images of figure 3.22(*d*) show the features observed by Wilson and Hirsch (1972). In particular, the black–white interfaces show characteristic angular fine structure, interpreted by Wilson and Hirsch as being due to partial loop dissociation. These bear a strong resemblance to the black–white images of SFT under this imaging condition (figure 3.17(a_1)),

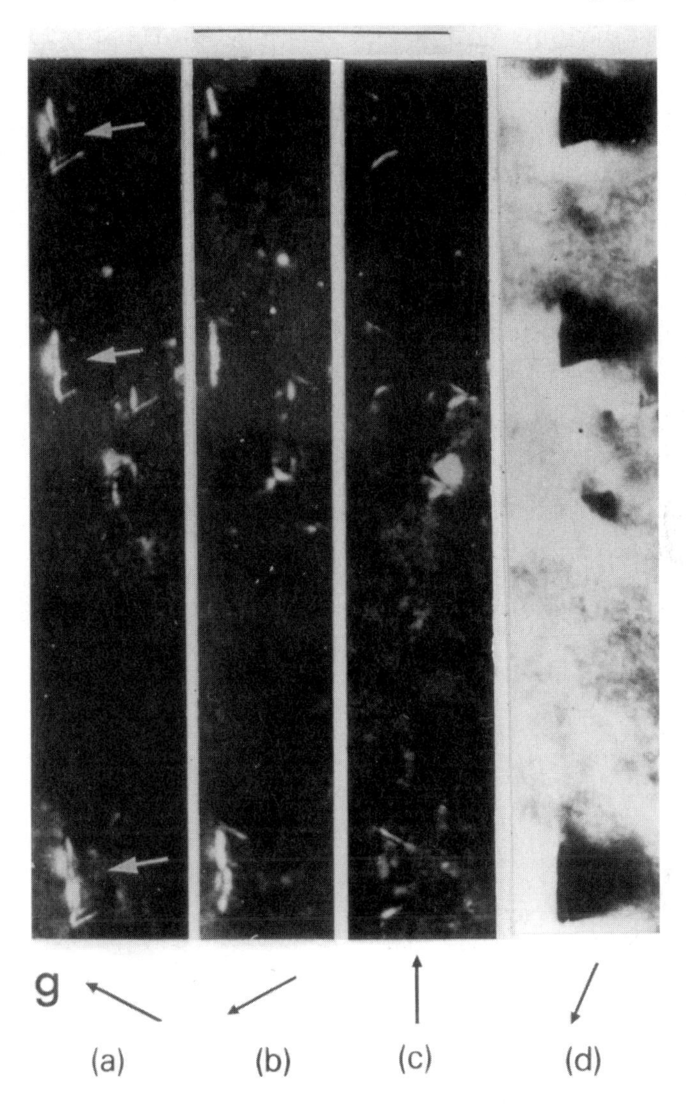

Figure 3.22. A contrast experiment on defects in heavy-ion irradiated silver-comparison with strong two-beam image: (*a*)–(*c*) weak-beam, and (*a*) $\boldsymbol{g} = 2\bar{2}0$, (*b*) $\boldsymbol{g} = 20\bar{2}$, (*c*) $\boldsymbol{g} = 02\bar{2}$, all with $s_g = 2 \times 10^{-1}$ nm^{-1} and the foil oriented close to [111]. (*d*) Dynamical dark field with $\boldsymbol{g} = 002$, with $s_g = 0$ and the foil orientated close to [110]. Scale marker 50 nm. From Jenkins (1974).

although for partially-dissociated loops the two straight segments of interface have unequal lengths.

- Defects in ion-irradiated copper were generally smaller than 5 nm and

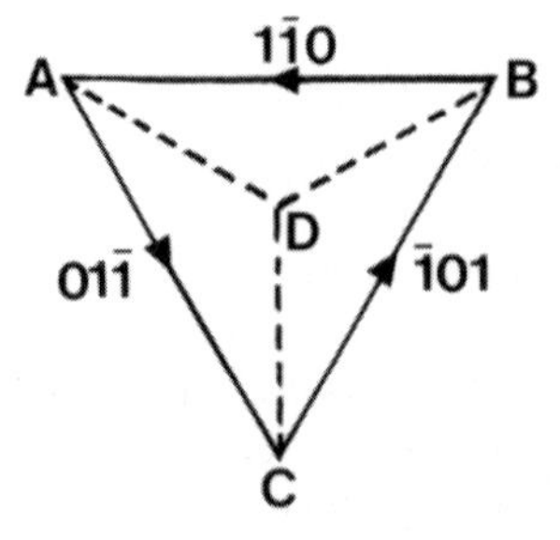

Reflection	Contrast type	Appearance of defect
2$\bar{2}$0	(a)-type	g
$\bar{2}$20	(b)-type	g
0$\bar{2}$2	(a)-type	g
02$\bar{2}$	(b)-type	g
±20$\bar{2}$	(c)-type	g

Figure 3.23. A schematic diagram summarizing the contrast changes of a single cluster under various weak-beam imaging conditions, such as is seen in figures 3.20–3.22. From Jenkins (1974).

systematic contrast experiments were not successful. However, the appearance of the defects was very similar to those in silver (see figure 3.24), suggesting that the defect geometries in the two cases are the same.

The postulated defect geometry deduced from these observations is shown in figure 3.25. It is assumed that the growing Frank loop, produced by the agglomeration of vacancies in the cascade depleted zone onto an inclined $\{111\}$ plane, is initially circular (diagram 1). As the local supersaturation of vacancies decreases, it becomes favourable for segments of the loop to line up along the $\langle 110\rangle$ directions to allow dissociation of the bounding Frank partial dislocations into stair-rod dislocations and Shockley partial dislocations on the intersecting $\{111\}$ planes (2). The sense of dissociation is influenced by the foil surface—dissociation towards the surface is favoured. As the loop accumulates more vacancies, pairs of corners B and A, C and D, E and F, move together (3). The final configuration is a partially-dissociated Frank loop, which may be thought of as a stage half-way between a Frank loop and an SFT.

The detailed contrast features seen in the micrographs are consistent with this model, and can again be understood from simple contrast considerations. We assume that the major part of the contrast comes from the inclined faces of stacking-fault ACEF, AFG and CEH, and from the Shockley partials GH and EJF. Let us assume that the faces of stacking-fault ACEF, AFG and CEH lie on the planes β, γ and α respectively, as shown in figure 3.25. Each $\{2\bar{2}0\}$ weak-beam reflection picks up stacking-fault contrast from two of the three inclined faces of fault. For example, in $\boldsymbol{g} = \pm 2\bar{2}0$, parallel to GA, the large area of stacking fault on β and the smaller area on α should both be in contrast. We would expect to see a long fringe, or fringes, parallel to $[01\bar{1}]$ (AC) and shorter fringes parallel to $[\bar{1}01]$ (CH). The change from (a)-type to (b)-type contrast can be understood qualitatively from the contrast expected from the Shockley partials GH and EJF, which both have $\boldsymbol{b} = \frac{1}{6}[2\bar{1}\bar{1}]$. For $\boldsymbol{g} = \pm 2\bar{2}0$, these dislocations have $|\boldsymbol{g} \cdot \boldsymbol{b} = 1|$. For $\boldsymbol{g} = \bar{2}20$ and $s_g > 0$, the image from the Shockley partial GH will lie on that side of the dislocation line nearer to the stair-rod AC. Similarly, the image of the Shockley EJF will lie on that side of the dislocation line nearer the stair-rod EF. We postulate that this leads to (a)-type contrast, where the stacking-fault fringes on β and α are partially masked by dislocation contrast. In $\boldsymbol{g} = 2\bar{2}0$ the images of these Shockley partial dislocations lie on the other side of the lines of the dislocations, in each case further away from the faces of fault on β and α. We expect this to lead to the simpler (b)-type contrast, where the stacking-fault fringes on these faces are seen clearly. Similar considerations apply for the other $\{2\bar{2}0\}$ reflections. For $\boldsymbol{g} = \pm\bar{2}02$, the faults on β and γ are in contrast, and the occurrence of (a)-type or (b)-type contrast again depends on the sign of $\boldsymbol{g}$. For $\boldsymbol{g} = \pm 02\bar{2}$, the two smaller of faces of fault on α and γ are imaged whilst the larger fault on β and the Shockleys GH and EJF are all out-of-contrast. This gives rise to (c)-type contrast. Note that (c) type contrast could easily be mistaken as arising from two separate, neighbouring clusters. This is a possible source of

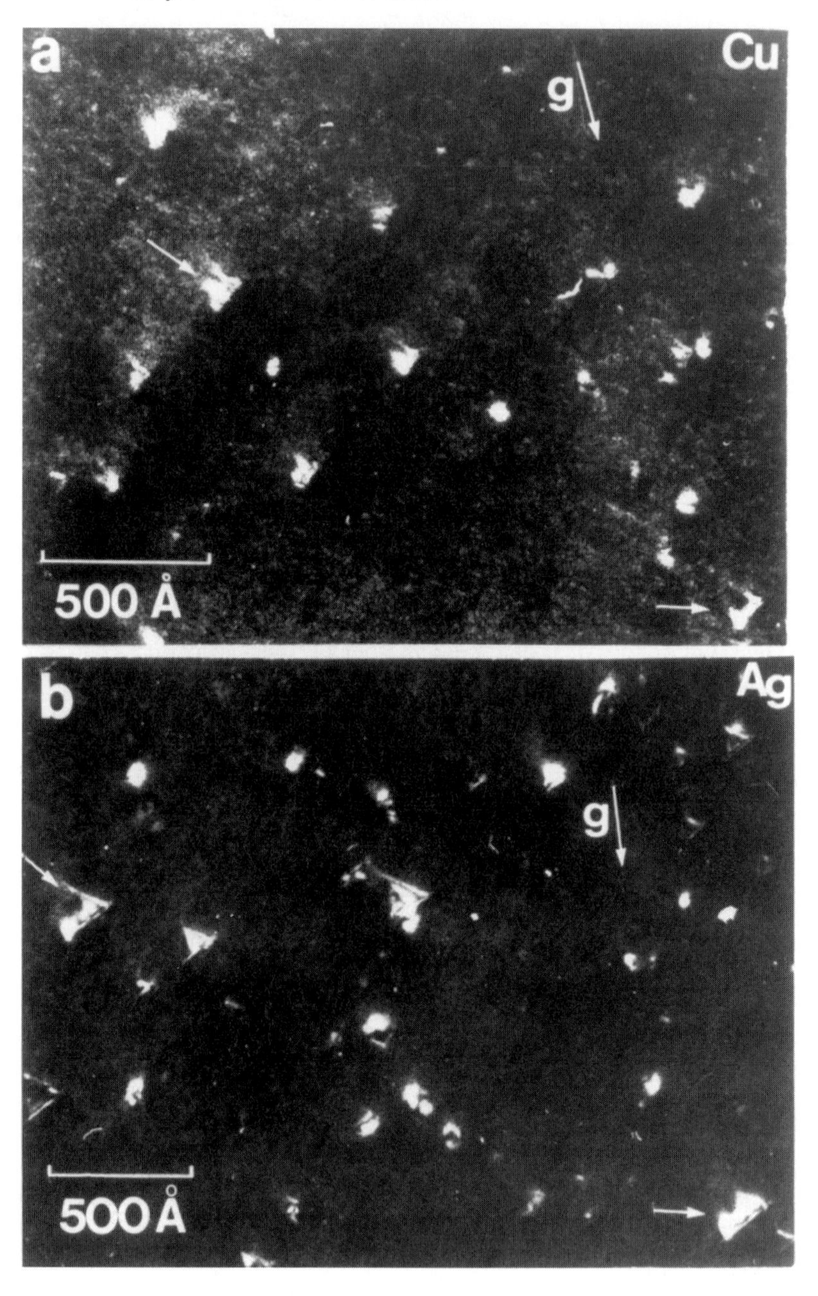

Figure 3.24. A comparison of defect clusters in (*a*) copper and (*b*) silver produced by heavy-ion irradiation. The imaging conditions are similar: $\boldsymbol{g} = 2\bar{2}0$, $s_g = 2 \times 10^{-1}\ \mathrm{nm}^{-1}$ and the foil oriented close to [111]. Although the defects in copper are smaller, they appear very similar to the clusters in silver. From Jenkins (1974).

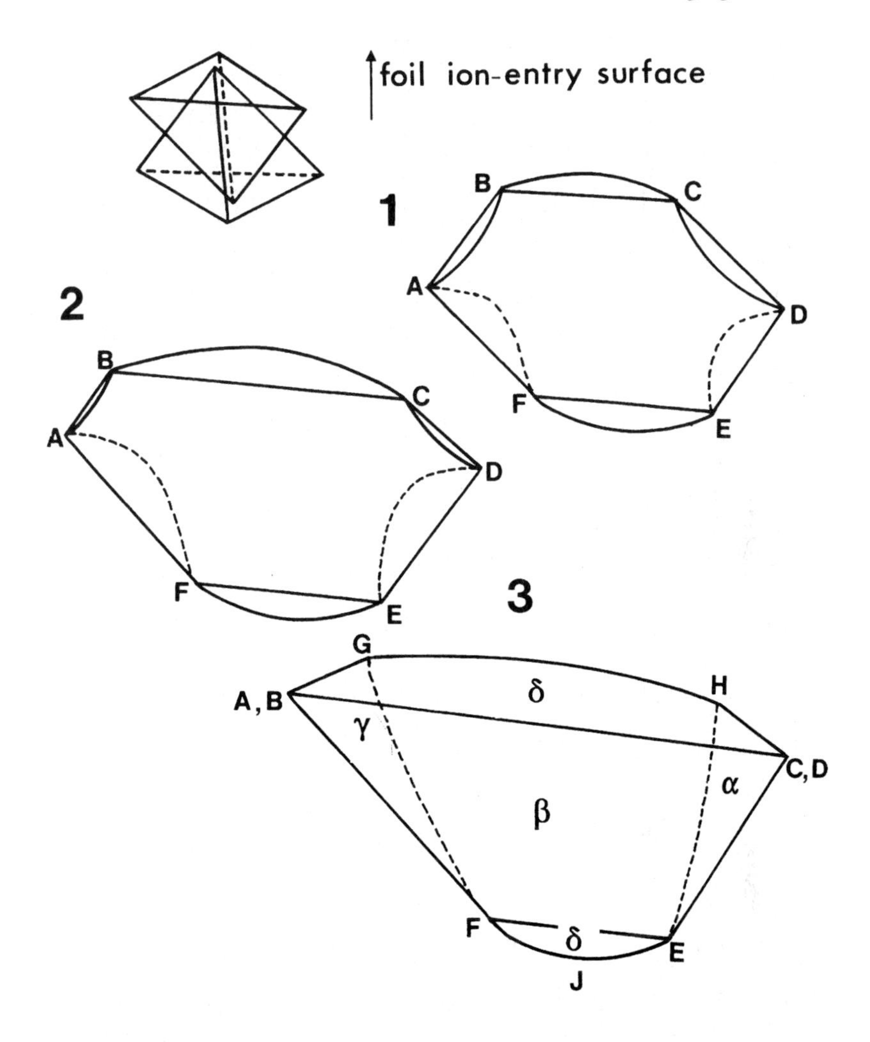

Figure 3.25. A schematic diagram of the defect geometry deduced from the contrast—a partially dissociated Frank loop. The various stages in the dissociation are discussed in the text. From Jenkins (1974).

errors in yield determinations using weak-beam microscopy. Note also that the dynamical two-beam images seen in figure 3.22(*d*) could be mistaken as arising from complete SFT.

3.2.2 Weak-beam analyses of clusters of size <5 nm

The defects in the previous examples were mostly of size >10 nm. Weak-beam microscopy has been used in many similar applications, and has been very successful in analysing defects of size greater than about 5 nm. The application of weak-beam microscopy to clusters smaller than 5 nm is less well established. Early authors (Häussermann *et al* 1973, Stathopoulos 1981) noted that the weak-beam contrast of small clusters was rather variable. Small defects generally appear as white dots, but the intensity and apparent size of these dots is sensitive to small changes in imaging conditions. In particular, for small defects with diameter $d < \xi_g^{\mathrm{eff}} \approx 5$ nm, image fine structure is often not seen, and it is usually not possible to follow systematic contrast changes. The intensity and extent of the contrast may be dependent on the foil thickness and defect depth. For this reason, these authors advised caution in the use of the weak-beam technique in studies of small clusters. However, in practice the technique is demonstrably superior to alternative techniques for identifying small clusters. Consider, for example, figure 2.5 which shows the same area imaged under a weak-beam condition and a down-axis condition. It is clear that many more defects are visible in the weak-beam micrograph.

We have recently made progress in developing weak-beam microscopy for defect counting and sizing for defects of size <5 nm. We have also shown that it is possible to determine the Burgers vectors of small (<5 nm) loops in neutron-irradiated iron, essentially by conventional contrast methods. It is necessary to record series of weak-beam micrographs with small changes in s_g, so that the sensitivity of the images to the exact diffracting conditions is recorded and analysed. This work is reported in more detail in sections 5.2.1 and 5.2.2.

So far, there has been no systematic simulation of weak-beam images of small dislocation loops. Such simulations would probably require both the two-beam and column approximations to be dropped.

3.3 The future for simulations of diffraction contrast images

The diffraction-contrast image simulations shown in section 3.1 have all employed the column approximation, and most have used the two-beam approximation. As has been seen, they have been very successful in predicting the general image features of small clusters imaged under dynamical conditions. As early as 1967, however, Howie and Basinski realized that experiments of the type described here do not test very deeply the dynamical theory of diffraction in imperfect crystals.

If experiments of a more quantitative kind were to be attempted, involving actual measurements of image intensity in an attempt to extract information about the detailed structure of defects, these approximations would have to be examined critically and perhaps discarded. In the intervening years there have been few critical experiments of this kind. For this reason there has been little incentive to improve the simulation of diffraction-contrast images. Much effort has gone into simulations of high-resolution images, using an alternative multi-slice development of dynamical theory, but the simulation of diffraction contrast

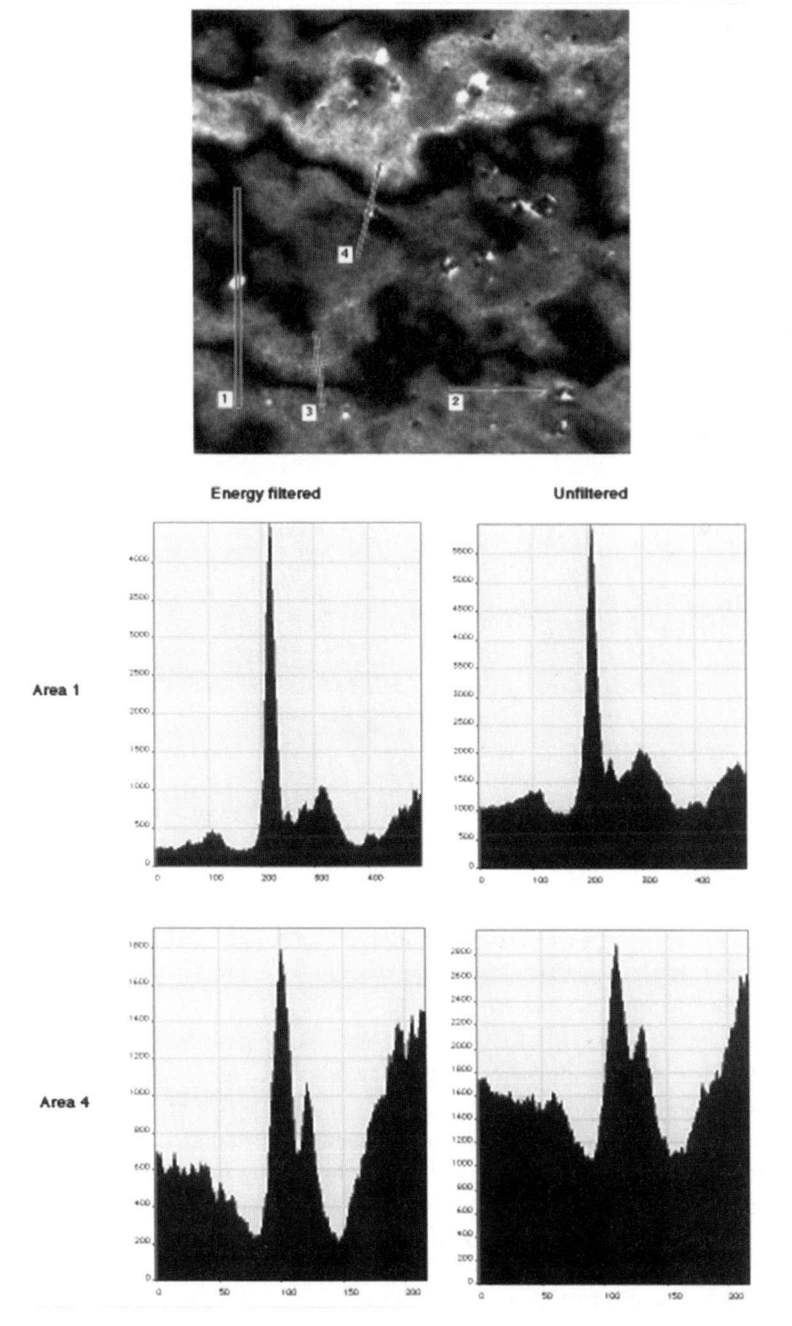

Figure 3.26. Improvement in the peak-to-background in weak-beam images obtained by energy-filtering.

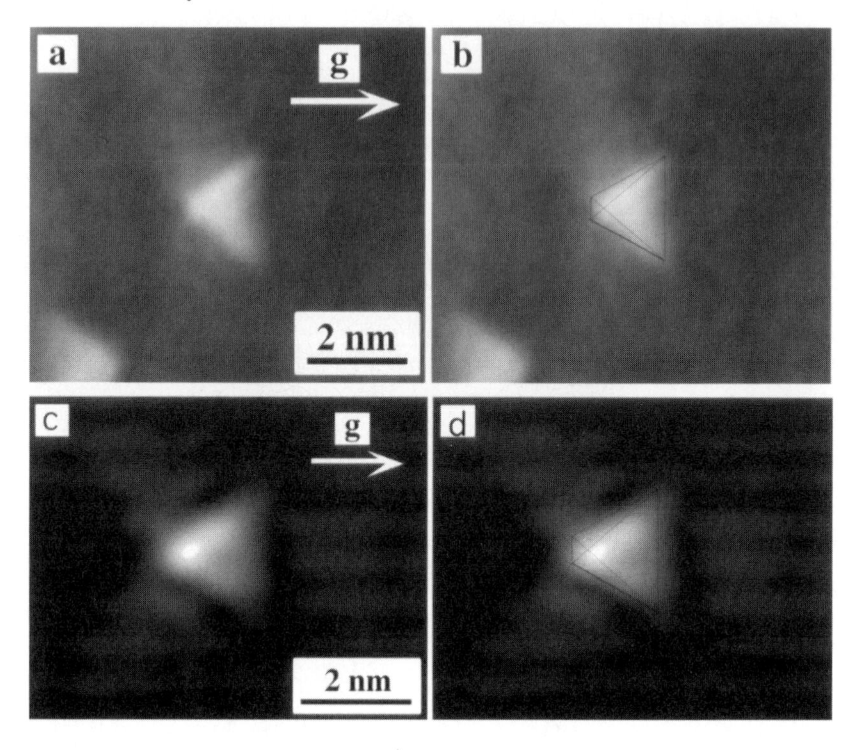

Figure 3.27. Comparison of experimental and computed images of SFT imaged under weak-beam conditions: (*a*) experimental image, $\boldsymbol{g} = 200$; (*b*) as (*a*) with the wire frame of an SFT superposed; (*c*) simulated image, $\boldsymbol{g} = 200$; (*d*) as (*c*) but with the wire frame of the molecular dynamics simulated SFT superposed. From Schäublin *et al* (1998).

images has not progressed and, in some ways, has even regressed. For example, some of the many-beam Bloch-wave codes available in the 1970s and 1980s seem to have been lost.

We anticipate that this situation will change. Already there is a pressing need for realistic weak-beam image simulations, where both the column and two-beam approximations are unlikely to be valid. For example, we have recently shown that weak-beam microscopy has good potential for finding and characterizing the elusive 'matrix damage' component of hardening in pressure vessel steels (see section 5.2). Developments such as the advent of energy-filtered images may make the identification of such defects possible, by detailed comparisons between experiment and theory of the type foreseen by Howie and Basinski. Preliminary experiments have shown that energy-filtering makes a significant improvement in peak-to-background ratios, which will greatly assist *quantitative* contrast measurements, see figure 3.26. It is likely in such comparisons that the displacement fields of small clusters will be derived from molecular dynamics simulations rather than elasticity theory. There has already been some progress in this area. Schäublin *et al* (1998) show images of SFT calculated in this way, using a multi-slice approach which avoids the column approximation (see figure 3.27). At present there are practical restrictions in the thickness of foil which can be included if the calculation is not to take up an exorbitant amount of processor time.

The quantitative evaluation of experiments of this kind will be aided by the recent development of charge coupled device (CCD) cameras which allow images and diffraction patterns to be captured digitally, see section 2.4. Previously these data had to be captured on photographic film, with the inherent difficulties of lack of linearity and sensitivity, slowness in recording images and long delays between data capture and analysis. CCD cameras are now used routinely for quantitative measurements of image and diffraction intensities over a wide dynamic range and for on-line analysis of data.

Chapter 4

Analysis of small centres of strain: determination of the vacancy or interstitial nature of small clusters

The fates of the point-defects produced by particle irradiation, mechanical stress or high temperatures is central to an understanding to the development of defect clusters and extended microstructure. The formation of dislocation loops and the interaction between migrating point-defects and these loops and existing dislocation microstructure can have profound effects on, for example, the mechanical properties of structural metal alloys. An understanding of these complex defect mechanisms is often based on a knowledge of the vacancy or interstitial nature of dislocation loops.

Whether dislocation loops or other clusters consist of agglomerates of vacancies or of interstitials is one of the most fundamental questions which can be asked. It is also one of the most difficult to answer, especially for smaller clusters.

We consider here first the situation for dislocation loops larger than about 10–20 nm, where the inside–outside technique provides a reliable recipe for loop nature determination as long as appropriate precautions are taken. Then we consider two techniques applicable to the determination of the nature of smaller loops: black–white stereo analysis and the $2\frac{1}{2}D$ technique. Both of these techniques have difficulties and the latter may be fatally flawed. Then we summarize techniques appropriate for determining the nature of SFT. Finally some indirect methods are briefly described.

4.1 The inside–outside method

The inside–outside technique is possibly the most reliable method of determining the nature of a loop, although in the case of non-edge loops its application is not completely straightforward. The technique relies on the change in apparent sizes of inclined dislocation loops in images obtained with $s_g \neq 0$ when the diffraction vector $\boldsymbol{g}$ is reversed. For this method to be applicable it is necessary for the dislocation line bounding the loop to be resolved so that, for example, a circular loop lying on an inclined plane images as an oval or double arc, at least in the 'outside' contrast condition. Dislocation image widths are of the order of $\frac{1}{3}\xi_g^{\rm eff}$. For conventional bright-field kinematical imaging, with s_g small and $w \ll 1$, $\frac{1}{3}\xi_g^{\rm eff}$ is typically $\sim$10 nm, while for weak-beam imaging with $s_g \approx 2 \times 10^{-1}$ nm^{-1}, $\frac{1}{3}\xi_g^{\rm eff} \sim 1.5$ nm. This limits the sizes of loops which can be analysed by this technique to $\geq$30 nm for bright-field kinematical imaging or about $\geq$5 nm for weak-beam imaging.

4.1.1 Edge loops

For simplicity we first consider the easier case of edge loops. The loop Burgers vectors are here defined using the Finish–Start/Right-hand (perfect crystal) (FS/RH) convention. We imagine viewing the loops from the *top*, electron-entry side of the foil[1]. The dislocation line sense is defined to be clockwise around the loop. A closed Burgers circuit is drawn around the dislocation in a sense related by a right-hand rule to the positive sense of the dislocation. The closure failure (from Finish to Start) in a similar circuit taken in a perfect crystal is the Burgers vector of the loop. The convention results in the sense of Burgers vector shown in figure 4.1. For edge interstitial loops $\boldsymbol{b}$ is parallel to the upward-drawn loop normal $\boldsymbol{n}$ (i.e. $\boldsymbol{b} \cdot \boldsymbol{n} > 0$ and $\boldsymbol{b} \cdot \boldsymbol{z} < 0$, where $\boldsymbol{z}$ is the beam direction, or '*down*'), whilst for edge vacancy loops $\boldsymbol{b}$ is antiparallel to $\boldsymbol{n}$ (i.e. $\boldsymbol{b} \cdot \boldsymbol{n} < 0$, and $\boldsymbol{b} \cdot \boldsymbol{z} > 0$).

The physical principle of the inside–outside technique is shown in figure 4.2. Edge dislocation loops of interstitial and of vacancy natures, lying on planes inclined to the electron beam, are imaged under bright-field kinematical or dark-field weak-beam conditions with $s_g > 0$ using the diffraction vectors shown. The regions where the local lattice rotations produced by the strain fields of the loops are in a sense to bring the reflecting planes back towards the Bragg condition are indicated by the curved arrows. These regions diffract strongly, and the image peak therefore lies inside or outside the projected position of the dislocation loop core: this is termed 'inside' contrast for the loops in figure 4.2(A) and (C). Changing the sign of $\boldsymbol{g}$ or (for weak-beam images) of s_g results in 'outside' contrast, as for the loops in figure 4.2(B) and (D). Figure 4.2 makes it clear that the inside or outside contrast behaviour of a given loop depends both on the loop nature—vacancy or interstitial—and the sense of inclination of the loop in the foil. Formally inside contrast arises when $(\boldsymbol{g} \cdot \boldsymbol{b})s_g < 0$, outside when $(\boldsymbol{g} \cdot \boldsymbol{b})s_g > 0$.

It is clear that in order to obtain the loop nature, the sense of inclination of the loop must be determined. This may be done by tilting the foil in a known sense, and noting the change in shape of the loop (see figures 4.8 and 4.9). In practice

[1] We use here the more common convention. In Hirsch *et al* (1977) the loop is viewed from beneath. This reverses the sense of $\boldsymbol{b}$, with consequent changes throughout the analysis.

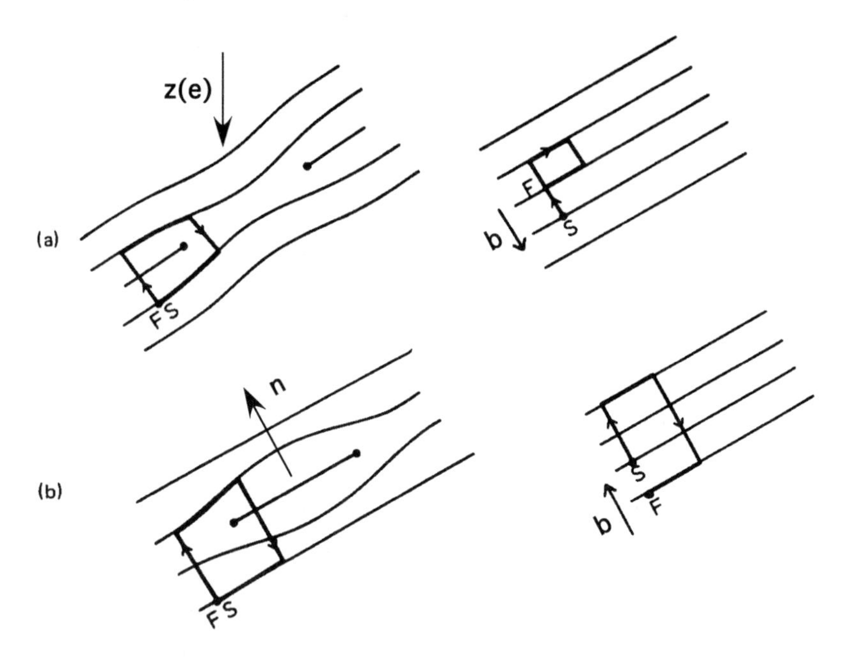

Figure 4.1. Defining the Burgers vector of inclined loops using the FS/RH (perfect crystal) convention, for (*a*) a vacancy loop, and (*b*) an interstitial loop. The dislocation line direction is taken as clockwise when viewed from above. A closed Burgers circuit is made around the dislocation line in the imperfect crystal. A similar circuit in the perfect material does not close. The Burgers vector is defined by the closure failure $\boldsymbol{b} = \boldsymbol{FS}$. Definitions of the loop normal $\boldsymbol{n}$ and electron beam direction z are also shown.

this may turn out to be difficult or tedious. In some cases, the sense of inclination can be found by an appeal to crystallographic arguments. For example, if the type of habit-plane is known (which is the case for the Frank loops considered here, which lie on $\{111\}$ planes), the sense of inclination may be established easily.

Let us now apply this analysis to determine the nature of the loops in silicon shown previously in figure 3.18. As discussed in chapter 3, the contrast experiment shows the majority of loops to be of the faulted Frank type, so they lie on $\{111\}$ planes with $\boldsymbol{b} = \frac{1}{3}\langle 111\rangle$. The steps are as follows:

- The diffraction vectors $\boldsymbol{g}$ were drawn on the micrographs using selected area diffraction patterns, which were recorded at the same time as the micrographs, together with a knowledge of the rotation calibration of the microscope. In order to do this, we needed to know the magnitude and sense

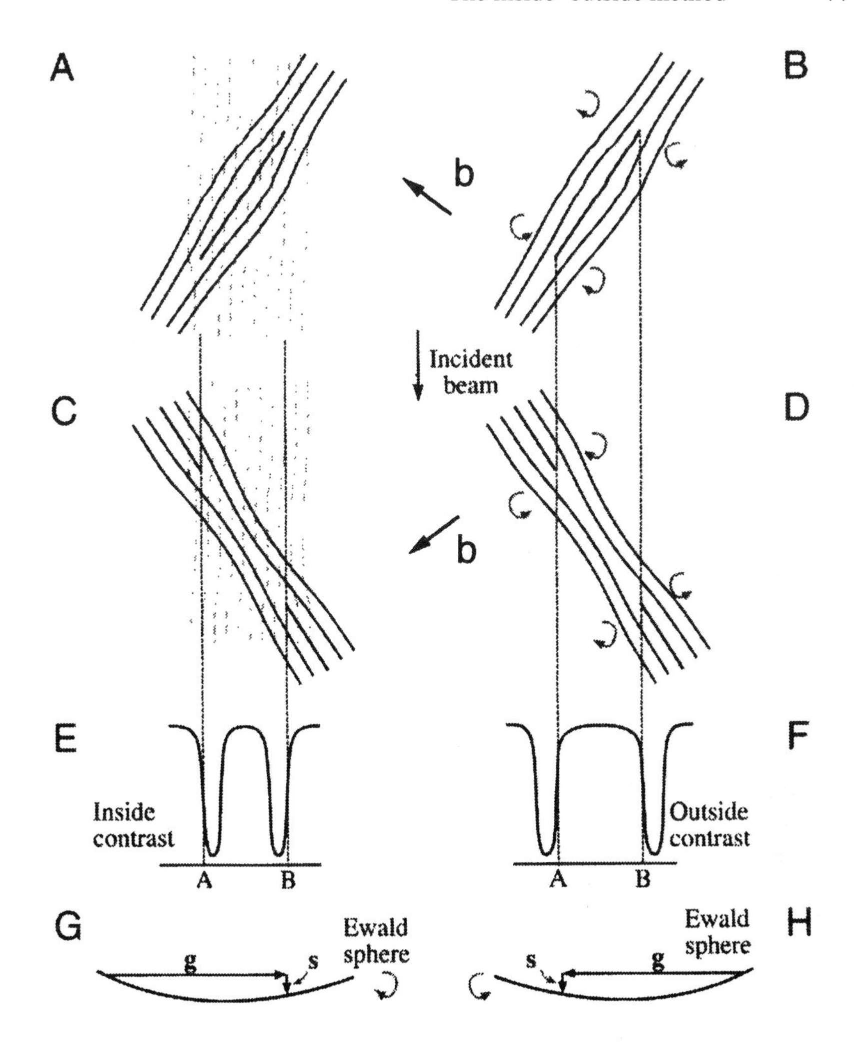

Figure 4.2. Schematic diagrams showing the principle of the inside–outside technique for loop nature determination. An inclined interstitial loop with diffracting planes (faint lines) is shown in (A) with the sense of plane rotations indicated in (B). The same is shown for a oppositely-inclined vacancy loop in (C) and (D). The position of the image contrast with respect to the projection of the loop dislocations is shown in (E) and (F) for the diffraction conditions shown in (G) and (H), respectively. In (H) the sign of $\boldsymbol{g}$ is reversed, but in both (G) and (H) s_g is positive. Inside contrast results when clockwise rotation of the diffracting planes brings them into the Bragg condition (G). Outside contrast results for the counterclockwise case (H). Everything is reversed if the loops are oppositely inclined with respect to the electron beam, or if s_g is negative. Formally, inside contrast arises when $(\boldsymbol{g} \cdot \boldsymbol{b})s_g < 0$, outside contrast when $(\boldsymbol{g} \cdot \boldsymbol{b})s_g > 0$. Figure largely based on figure 25.9 of Williams and Carter (1996).

of the rotation angle ϕ between diffraction pattern and image and the sense of $\boldsymbol{g}$[2].

- It is of course necessary that there be internal self-consistency in the indexing of all planes and directions. We start the indexing by taking the foil plane as (111) and the beam direction (down in the microscope) as $\boldsymbol{z} = [111]$.
- We now assign indices to the diffraction vector $\boldsymbol{g}$ in figure 3.18(a), which is of the type $\{\bar{2}20\}$. Since there is three-fold symmetry at the [111] orientation, there is some arbitrariness in the choice. The reflections $(\bar{2}20)$, $(20\bar{2})$ and $(0\bar{2}2)$ are equivalent. The reflections $(2\bar{2}0)$, $(\bar{2}02)$, and $(02\bar{2})$ are also equivalent. We choose to label $\boldsymbol{g}$ in figure 3.18(a) as either $\boldsymbol{g} = \bar{2}20$ or $\boldsymbol{g} = 2\bar{2}0$. The choice is not arbitrary i.e. $\boldsymbol{g} = \bar{2}20$ and $\boldsymbol{g} = 2\bar{2}0$ are not equivalent. By tilting the specimen whilst observing the diffraction pattern the positions of the $\langle 211 \rangle$ poles relative to the [111] pole were found, and by referring to a Kikuchi map (figure 4.3) or Thompson tetrahedron (see the appendix) it was established that the correct choice is $\boldsymbol{g} = \bar{2}20$.
- Further indexing is most easily achieved by the use of a Thompson tetrahedron, which can now be placed on the micrograph. The negatives were printed with the emulsion-side *down* in the enlarger, so the figure shows the loops as they would appear if viewed from below the specimen. The beam direction $\boldsymbol{z} = [111]$ (*down* in the microscope) therefore points *out* of the plane of the paper. We therefore place the Thompson tetrahedron with its apex D *below* the plane of the paper, and its edge $\boldsymbol{AB} = \bar{1}10$ aligned with $\boldsymbol{g}$. All directions and planes can now be assigned consistent indices.
- In figure 3.18(a)–(c), the Frank loops on planes α and β can be recognized by inspection from their projected shapes. These loops show inside–outside contrast changes if the sign of either $\boldsymbol{g}$ or of s_g is changed. We note that loops on α (marked A) show outside contrast in figure 3.18(a) and (c), and inside contrast in figure 3.18(b), while the loops on β (marked B) show the reverse.
- The Burgers vector of Frank loops on α in figure 3.18 is $\boldsymbol{b} = \pm\frac{1}{3}[\bar{1}1\bar{1}]$. In figure 3.18($a$), $\boldsymbol{g} = \bar{2}20$ and s_g is positive, and the loops on α are in outside contrast. Outside contrast implies $(\boldsymbol{g} \cdot \boldsymbol{b})s_g > 0$, and so $\boldsymbol{b} = \frac{1}{3}[\bar{1}1\bar{1}]$ rather than $\boldsymbol{b} = \frac{1}{3}[1\bar{1}1]$. $\boldsymbol{b}$ is therefore parallel to the *upward*-drawn loop-plane normal $\boldsymbol{n} = [\bar{1}1\bar{1}]$, with $\boldsymbol{b} \cdot \boldsymbol{z} < 0$. The loops are therefore *interstitial* in nature. The same conclusion may be reached by analysing the inside–outside contrast behaviour of loops on the other two inclined $\{111\}$ planes β and γ.

[2] ϕ can be calibrated by recording micrographs and superimposed diffraction patterns of molybdenum trioxide crystals (see Hirsch *et al* (1977): 13). In applying the rotation calibration, care must be taken to note whether the micrographs were printed with the emulsion-side down or up, so that rotations are made in the correct sense. In order to determine unambiguously the sense of $\boldsymbol{g}$—that is, in which direction the arrow points—the possibility of image inversions has to be considered. A straightforward way of determining the sense of $\boldsymbol{g}$ which avoids problems associated with image inversions is described in the book by Loretto (1984: 42). This involves defocusing an image of the diffraction pattern in a known sense.

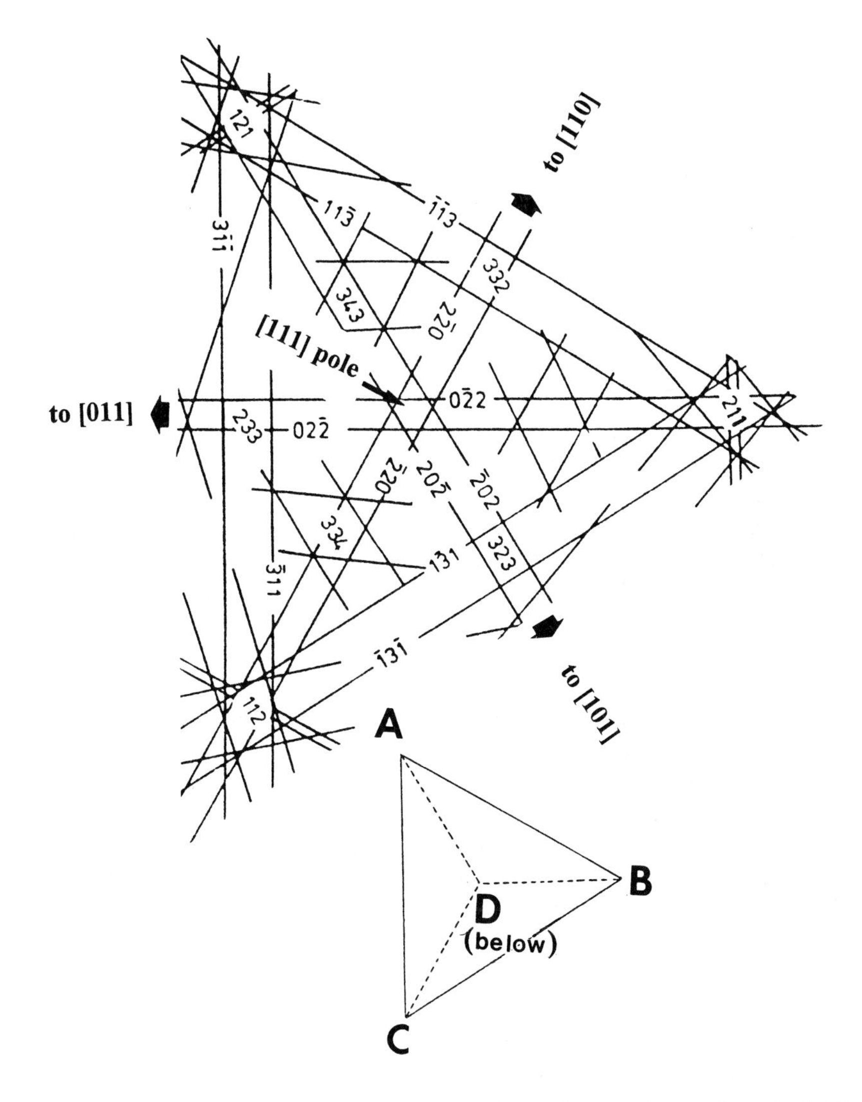

Figure 4.3. Schematic diagram of Kikuchi map for an fcc crystal centred on the [111] pole. Specimen tilting is used to establish the positions of the three ⟨211⟩ poles shown. The $\boldsymbol{g}$ vectors are then labelled as indicated. The Kikuchi map is viewed with $z = [111]$ out of the plane of the paper, as it would appear if viewed from below the specimen (as on a print made with the negative emulsion down in the enlarger).

For Frank loops, analysis of the loop nature is straightforward because the loop habit-plane is determined if the 'axis' of $\boldsymbol{b}$ is known. The 'axis' means the direction of $\boldsymbol{b}$ without distinction between positive or negative directions, and is found easily by the use of the $\boldsymbol{g} \cdot \boldsymbol{b} = 0$ invisibility criterion. For an edge loop, determination of the axis of $\boldsymbol{b}$ is clearly equivalent to finding the loop habit-plane, since $\boldsymbol{b}$ is parallel or antiparallel to $\boldsymbol{n}$. If the habit-plane is known, determination of the sense of inclination of the loop is relatively trivial. Complications arise, however, if the loops are not of pure edge character and these will now be discussed. Much of the following discussion is based on the short paper by Carpenter (1976).

4.1.2 Non-edge loops

Most, but not all, loops encountered in practice will show the same inside–outside contrast behaviour as an edge loop with the same Burgers vector. If non-edge loops are encountered, certain orientations show the reverse inside–outside behaviour compared with edge loops with the same Burgers vector. The examples shown schematically in figure 4.4 demonstrate the expected inside–outside contrast behaviour of a non-edge interstitial loop in various orientations, with the axis of $\boldsymbol{b}$ fixed in each case.

(*a*) The loop shows the same inside–outside contrast behaviour as an edge loop. This is similar to the case discussed previously. To determine its nature, it is sufficient to know the inside–outside behaviour, and either the sense of inclination of the loop or the axis of its Burgers vector.

(*b*) The loop shows non-edge behaviour with respect to the Burgers vector $\boldsymbol{b}$. We can imagine attaining this configuration by rotating the loop shown in (*a*) clockwise on its glide cylinder through the edge-on orientation $\boldsymbol{n} \cdot \boldsymbol{z} = 0$. For orientations beyond this, the FS/RH convention leads to a reversal in the sense of $\boldsymbol{b}$.

(*c*) The loop shows non-edge behaviour with respect to inclination. We can imagine attaining this configuration by further clockwise rotation of the loop through the pure shear orientation $\boldsymbol{n} \cdot \boldsymbol{b} = 0$, if both $\boldsymbol{n}$ and $\boldsymbol{b}$ lie in the plane containing $\boldsymbol{g}$ and $\boldsymbol{z}$, or more generally $\boldsymbol{n}_\mathrm{p} \cdot \boldsymbol{b}_\mathrm{p} = 0$. Since $\boldsymbol{g} \cdot \boldsymbol{b}$ is again positive, we again have edge-type behaviour with respect to $\boldsymbol{b}$.

It is clear from the foregoing discussion that a loop may be *fully* characterized by *all* of the following:

(a) noting the inside–outside behaviour in $\pm\boldsymbol{g}$;

(b) finding the 'axis' of $\boldsymbol{b}$ (i.e. by performing a $\boldsymbol{g} \cdot \boldsymbol{b} = 0$ contrast experiment); and

(c) finding the loop normal $\boldsymbol{n}$ explicitly.

Clearly if *all* of this information is available the loop nature can be determined unambiguously and there is no question of 'unsafe' orientations. Unsafe orientations (which are defined later) only arise when, to simplify the experimental procedure, all of this information is not obtained for a single loop. For example, we could proceed as in method (i) described in the following by

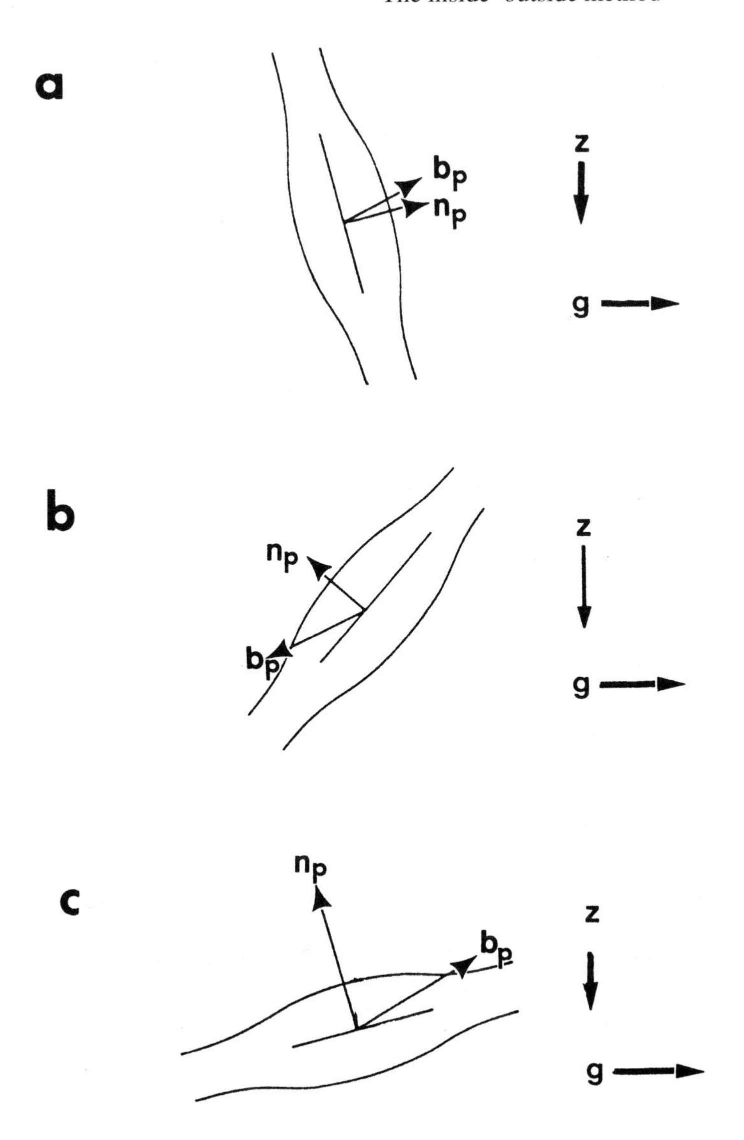

Figure 4.4. The effect of the loop orientation on the inside–outside contrast behaviour, illustrating when a loop with a shear component does, or does not, show the same contrast behaviour as an edge loop with the same Burgers vector. An inclined non-edge interstitial loop is shown in various orientations. The $\boldsymbol{g}$-vector is as shown. $\boldsymbol{b}_{\mathrm{p}}$ and $\boldsymbol{n}_{\mathrm{p}}$ are the projections of the Burgers vector $\boldsymbol{b}$ and loop normal $\boldsymbol{n}$ in the plane containing z and $\boldsymbol{g}$. The loop shows: (*a*) the same inside–outside contrast behaviour as a pure edge loop, i.e. outside contrast if $s_g > 0$; (*b*) non-edge behaviour with respect to $\boldsymbol{b}$, inside contrast if $s_g > 0$, (*c*) non-edge behaviour with respect to inclination, outside contrast if $s_g > 0$. From Maher and Eyre (1971).

determining only the axis of $\boldsymbol{b}$, and not the sense of inclination of the loop. In this case, the assumption of edge behaviour would lead to the wrong conclusion for the nature of loops in orientation (*b*) in figure 4.4.

Various techniques, which use partial information to determine loop natures, are described in the literature including the following ones.

Method (i) $\pm\boldsymbol{g}$ *pair plus the axis of* $\boldsymbol{b}$

In this method, which was introduced by Maher and Eyre (1971), a contrast experiment is performed to find which Burgers vectors are present, but the sense of inclination of the loops is not determined. Families of loops with a particular Burgers vector are 'unsafe' and an analysis of their nature should be avoided, if the possibility exists that some of the loops have the orientation shown in figure 4.4(*b*). Put formally, a nature analysis should not be attempted if $\boldsymbol{n}$ could lie outside the region of a stereograph containing the beam direction $\boldsymbol{z}$ and bounded by the lines $\boldsymbol{n} \cdot \boldsymbol{b} = 0$ and $\boldsymbol{n} \cdot \boldsymbol{z} = 0$.

The analysis proceeds by making the assumption that it is possible to specify an upper limit ϕ_{m} to the acute angle ϕ between $\boldsymbol{b}$ and the loop normal $\boldsymbol{n}$. It is then possible to select a crystal orientation $\boldsymbol{z}$, where the acute angle α between $\boldsymbol{b}$ and $\boldsymbol{z}$ is sufficiently small that the loop exhibits the same contrast as an edge loop of the same Burgers vector, irrespective of its inclination. Specifically, it is necessary to avoid attempting to analyse loops for which the angle α between $\boldsymbol{b}$ and $\boldsymbol{z}$ is $>(90° - \phi_{\mathrm{m}})$.

The method is probably illustrated most clearly by considering an example. We shall consider an analysis of the most common loops found in irradiated bcc crystals, namely prismatic loops with Burgers vectors of type $\boldsymbol{b} = \frac{1}{2}\langle 111 \rangle$. It is generally accepted that loops in bcc crystals nucleate as faulted loops on the closest-packed planes of type $\{110\}$. These faulted loops unfault to prismatic loops by the Eyre–Bullough mechanism, and subsequently may or may not rotate on their glide cylinders towards the edge orientation in order to lower their line length. It follows that ϕ_{m} corresponds to loops which have retained their original $\{110\}$ habit-plane, giving a value of $\phi_{\mathrm{m}} = 35°$. Loop variants with a particular Burgers vector may be safely analysed at those orientations $\boldsymbol{z}$ for which $\alpha < (90° - 35°) = 55°$. If $\alpha > 55°$ the orientation is unsafe for that loop variant.

Consider now specifically the analysis of $\frac{1}{2}\langle 111 \rangle$ loops performed using $\boldsymbol{g} = 200$ at a foil orientation $\boldsymbol{z} = [012]$. The possible $\frac{1}{2}\langle 111 \rangle$ type loop variants have Burgers vectors $\pm\frac{1}{2}[1\bar{1}1]$, $\frac{1}{2}[111]$, $\frac{1}{2}[11\bar{1}]$ and $\frac{1}{2}[\bar{1}11]$. For $\boldsymbol{b} = \pm\frac{1}{2}[\bar{1}11]$ and $\pm\frac{1}{2}[111]$, the angle $\alpha = 39.2°$, so loops of these types are safe; all would be expected to show the same inside–outside behaviour as pure edge loops with these Burgers vectors.

However, for $\boldsymbol{b} = \pm\frac{1}{2}[11\bar{1}]$ and $\boldsymbol{b} = \pm\frac{1}{2}[1\bar{1}1]$, $\alpha = 75°$, and so loops with these Burgers vectors are unsafe; the possibility exists that they may show reverse contrast.

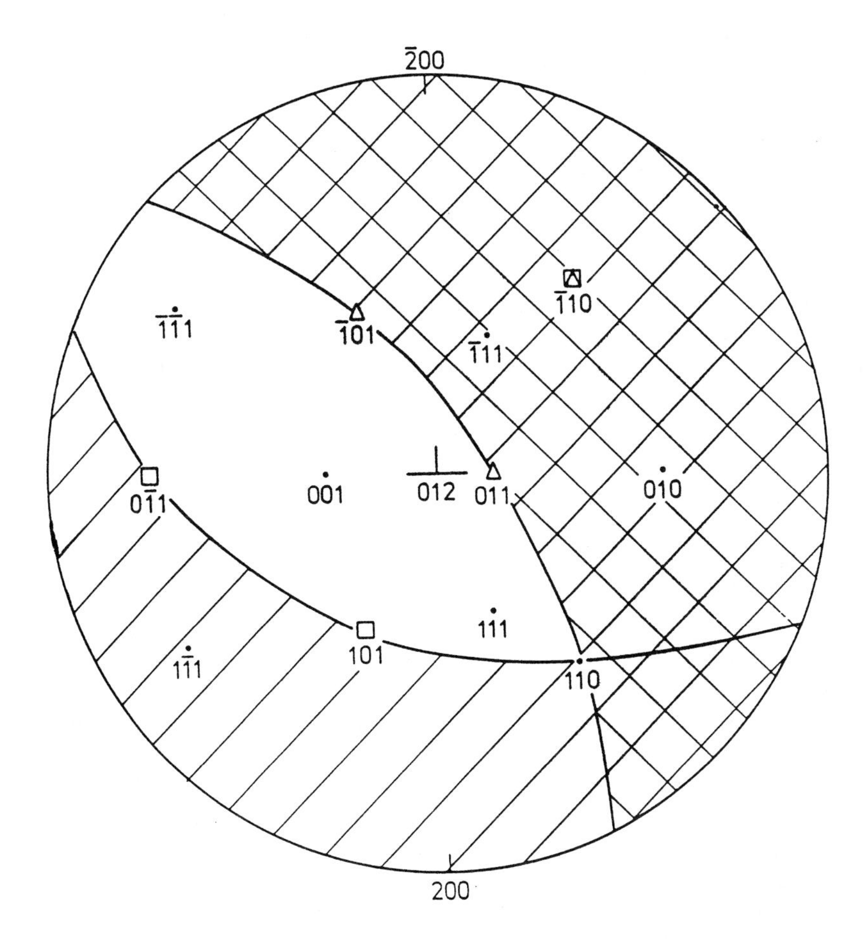

Figure 4.5. A stereogram defining the conditions which must be fulfilled for a non-edge loop to show the same contrast as a pure edge loop of the same Burgers vector for the analysis condition described in the text. The hatched areas define orientations of the loop normal $\boldsymbol{n}$ where a non-edge loop would show the opposite contrast to a pure edge loop of the same Burgers vector $\boldsymbol{b} = \pm\frac{1}{2}[\bar{1}11]$. The cross-hatched area defines the similar orientations for $\boldsymbol{b} = \pm\frac{1}{2}[1\bar{1}1]$. Similar regions could also be defined for the other two loop variants but are not shown for clarity.

This procedure is illustrated in the stereogram of figure 4.5. The constraint $\boldsymbol{n} \cdot \boldsymbol{z} = 0$ is the outer circle of the stereogram. The $\boldsymbol{n} \cdot \boldsymbol{b} = 0$ constraint for loops with $\boldsymbol{b} = \pm\frac{1}{2}[\bar{1}11]$ (one of the safe variants) and $\boldsymbol{b} = \pm\frac{1}{2}[1\bar{1}1]$ (an unsafe variant) are shown, and the respective regions of reverse contrast are hatched and cross-hatched respectively. The possible loop nucleation planes for loops with these Burgers vectors are shown as triangles (△) and squares (□) respectively.

None of the nucleation planes for the $\boldsymbol{b} = \pm\frac{1}{2}[\bar{1}1\bar{1}]$ loops lies within the hatched region, so this orientation is safe for loops with this Burgers vector. All loops with $\boldsymbol{b} = \pm\frac{1}{2}[\bar{1}\bar{1}1]$ will show edge-type contrast. However, the $(\bar{1}10)$ nucleation plane for the $\boldsymbol{b} = \pm\frac{1}{2}[1\bar{1}\bar{1}]$ loop lies within the cross-hatched region. Loops of this Burgers vector with habit-planes close to this nucleation plane will show reverse contrast. Loops with the same Burgers vector which nucleated on the other two possible $\{110\}$ planes will show edge-type contrast. If we do not find the loop habit-plane explicitly, we therefore have no way of knowing whether a given loop with $\boldsymbol{b} = \pm\frac{1}{2}[1\bar{1}\bar{1}]$ will show edge-type or reverse contrast. The [012] orientation is therefore unsafe for analysing loops with this Burgers vector. Clearly exactly the same arguments apply to the other safe and unsafe variants respectively.

This specific example also demonstrates a more general 'rule of thumb' for the use of this method in a material of any type. Since most loops encountered in practice are likely to have appreciable edge components, one should not attempt to analyse loops which are close to the edge-on orientation.

Method (ii) $\pm\boldsymbol{g}$ pair plus determination of $\boldsymbol{n}$

In this method the Burgers vectors are not determined, but the habit-planes of the loops are found explicitly. Because it is usually easier to determine $\boldsymbol{b}$ than $\boldsymbol{n}$, this technique is not used so frequently as the Maher and Eyre method (i) and so it will not be considered in the same detail. The loop normal $\boldsymbol{n}$ may be determined by tilting the loop to an edge-on configuration. A less difficult but also less precise method may be to monitor the change in projected shape of the loop (see section 4.1.4 and figures 4.8 and 4.9).

The unsafe orientation is shown in figure 4.4(*c*). The corresponding unsafe loop normals are bounded by the conditions $\boldsymbol{n}_\mathrm{p} \cdot \boldsymbol{b}_\mathrm{p} = 0$ (where $\boldsymbol{n}_\mathrm{p}$ and $\boldsymbol{b}_\mathrm{p}$ are the projections of $\boldsymbol{n}$ and $\boldsymbol{b}$ on the plane containing $\boldsymbol{g}$ and $\boldsymbol{z}$) and $\boldsymbol{n} \cdot \boldsymbol{g} = 0$. Again the analysis proceeds by making the assumption that it is possible to specify an upper limit ϕ_m to the acute angle ϕ between $\boldsymbol{b}$ and $\boldsymbol{n}$. It is necessary to avoid loop orientations for which the acute angle between $\boldsymbol{z}$ and either $\boldsymbol{n}_\mathrm{p}$ or $\boldsymbol{b}_\mathrm{p}$ is less than ϕ_m.

Provided loops are roughly circular, are fairly steeply inclined and have appreciable edge character, it is sufficient to determine only the sense of inclination (which is essentially the method of Edmondson and Williamson (1964)). However, the possibility of obtaining reverse contrast in unsafe reflections must be recognized. See Carpenter (1976) for a more detailed discussion.

Method (iii) $\pm\boldsymbol{g}$ pair plus sense of inclination and axis of $\boldsymbol{b}$

This information should suffice for a characterization of most loops. Knowledge of the axis of $\boldsymbol{b}$ allows the regions of reverse contrast to be specified as in method (i). Finding the sense of inclination, together with the known value of ϕ_m, usually allows one to predict whether pure-edge or reverse contrast is expected for any particular loop. Ambiguity arises only for near-shear loops, which are rarely encountered in practice, or for certain orientations which are in any case avoided because they give poorly defined inside–outside contrast (specifically when the angle between $\boldsymbol{b}$ and $\boldsymbol{b}_{\mathrm{p}}$ or between $\boldsymbol{n}$ and $\boldsymbol{n}_{\mathrm{p}}$ approaches 90°).

This method is essentially equivalent to the technique of Föll and Wilkens (1975), although these authors use a different definition of $\boldsymbol{b}$ which somewhat simplifies the analysis by avoiding the need to define unsafe orientations (although as Carpenter points out, this redefinition is not essential).

4.1.3 Example: loops in neutron-irradiated iron

A practical example of loop nature determination for a loop in neutron-irradiated iron taken from the work of Robertson (1982) is illustrated in figures 4.6–4.9. Figure 4.6 shows the Burgers vector analysis of a loop in an (001) foil. The loop is clearly visible in the micrograph of figure 4.6(*a*) obtained using the diffraction vector $\boldsymbol{g} = 1\bar{1}0$, but is out of contrast (i.e. $\boldsymbol{g} \cdot \boldsymbol{b} = 0$) in figures 4.6(*b*) and (*c*) obtained using $\boldsymbol{g} = 110$ and $\boldsymbol{g} = 21\bar{1}$ respectively. This establishes the Burgers vector as $\boldsymbol{b} = \pm\frac{1}{2}[1\bar{1}1]$. The inside–outside experiment is shown in figure 4.7. All micrographs were obtained under kinematical conditions with $s_g > 0$. It is easily confirmed that both of the foil orientations used are safe for this loop variant (see method (i)) and so the loop will show edge-type behaviour. The loop shows outside contrast in figures 4.7(*b*), (*d*) and (*f*), corresponding to $(\boldsymbol{g} \cdot \boldsymbol{b})s_g > 0$, so that $\boldsymbol{b} = +\frac{1}{2}[1\bar{1}1]$. The indexing of diffraction vectors is consistent with [001] pointing into the plane of the paper. Since these are emulsion-down prints, which view the specimen as if from below, this is the direction of the upward-drawn foil normal. The electron beam direction is therefore $\boldsymbol{z} = [00\bar{1}]$ (see figure 4.1). Since $\boldsymbol{b} \cdot \boldsymbol{z} < 0$ the loop is interstitial in nature. Note that there is some redundancy here. One inside–outside pair would have sufficed in principle, but the result is perhaps more convincing if seen consistently in several reflections.

4.1.4 Determining the loop habit-plane

The only assumption in the previous analysis, that the maximum angle between $\boldsymbol{b}$ and $\boldsymbol{n}$ is 35° (see method (i)), may be removed by finding $\boldsymbol{n}$ explicitly. The techniques for doing this and their range of applicability have been discussed by Maher and Eyre (1971). Two general ways are available:

(a) The loop may be tilted to an edge-on orientation and $\boldsymbol{n}$ evaluated from the loop trace.
(b) Observation of the change in projected direction of the loop major axis on tilting.

Method (a) was not possible in the present case because of the limited specimen tilting available on the stage used. Method (b) was therefore employed and is illustrated in figures 4.8 and 4.9. It is best to choose two orthogonal axes

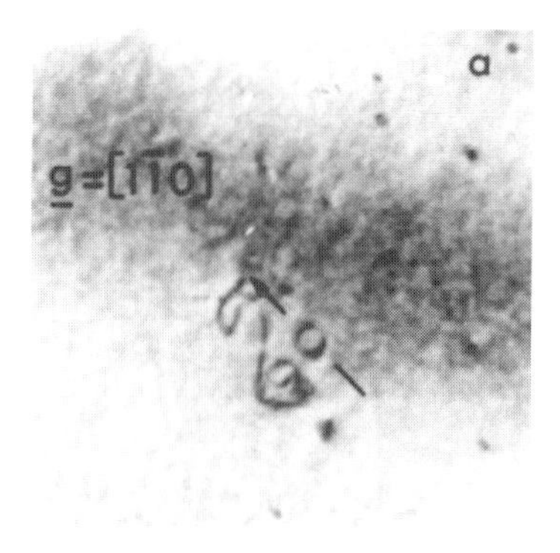

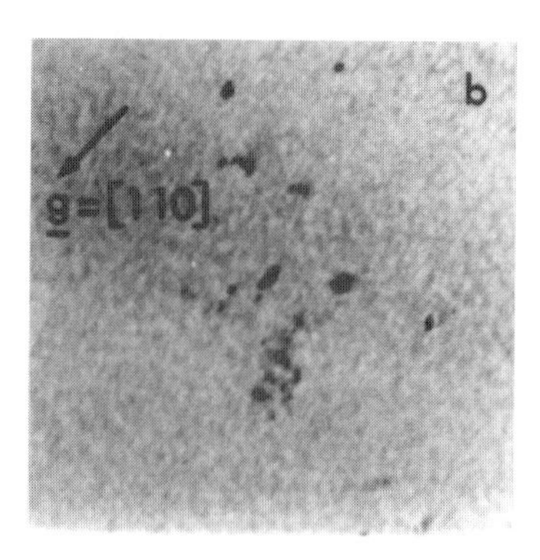

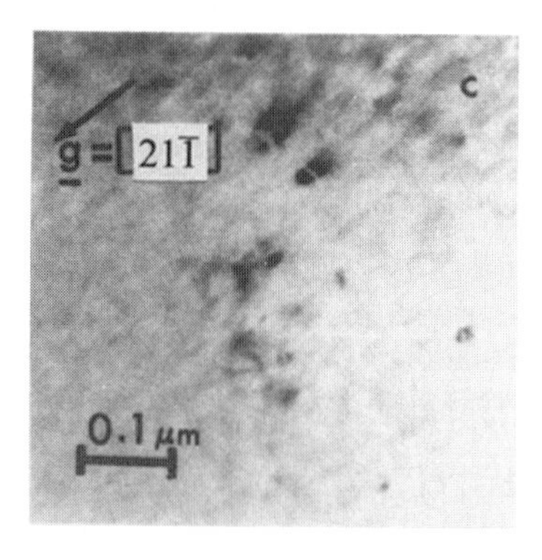

Figure 4.6. Determination of the Burgers vector of a loop using the $|\boldsymbol{g} \cdot \boldsymbol{b}| = 0$ invisibility criterion. In (*a*) and (*b*) the beam direction is close to [001], and in (*c*) it is close to [113]. The $\boldsymbol{g}$-vectors are as shown. In (*b*) and (*c*), $\boldsymbol{g} \cdot \boldsymbol{b} = 0$. The loop Burgers vector is therefore $\boldsymbol{b} = \pm\frac{1}{2}[1\bar{1}1]$. Micrographs here and in the next three figures courtesy of Robertson (1982).

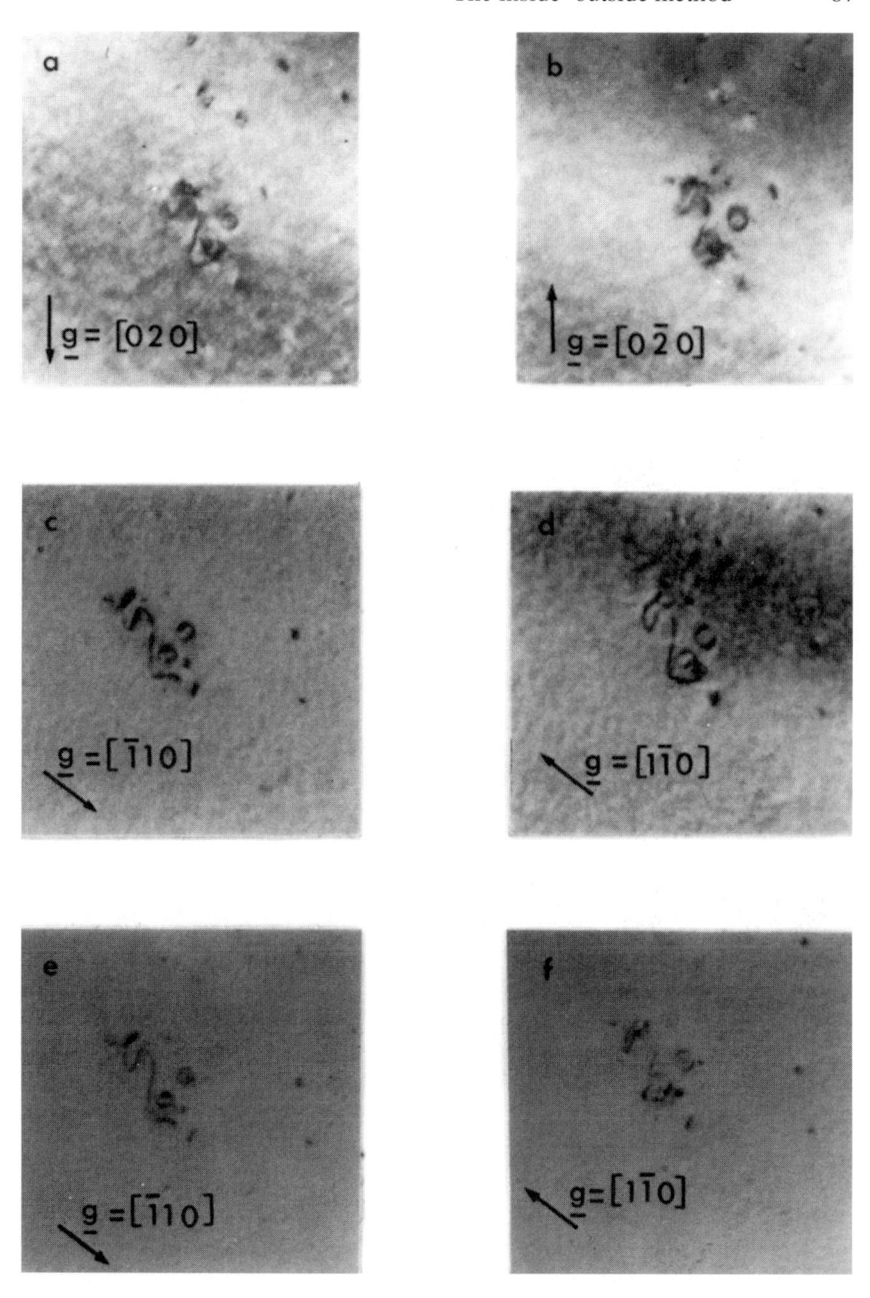

Figure 4.7. Determination of the nature of the loop shown in figure 4.6 by the method of Maher and Eyre (1971). In (*a*)–(*d*) the electron-beam direction is close to [001], and in (*e*) and (*f*) it is close to [115]. The loop shows outside contrast in (*b*), (*d*) and (*f*). It is therefore interstitial in nature (see text).

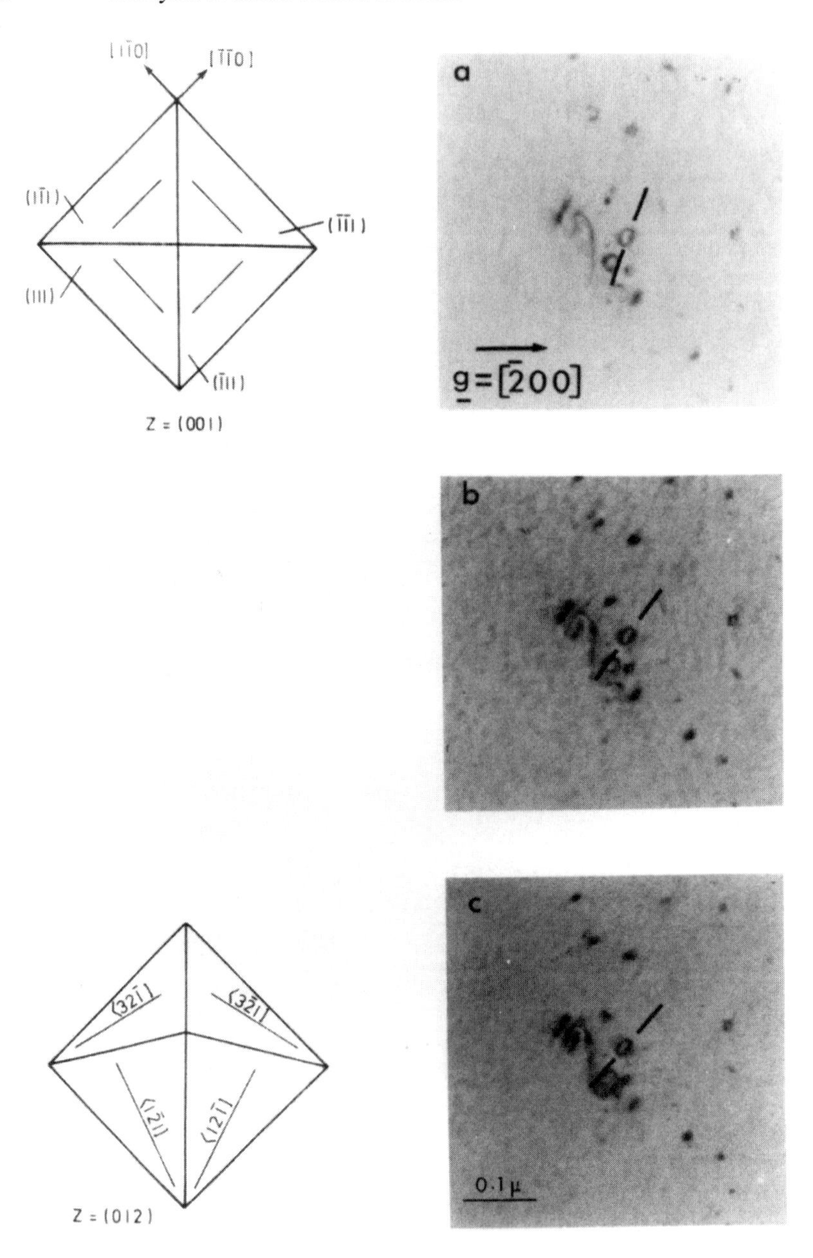

Figure 4.8. Tilting experiments to determine the habit-plane of the loop shown in figures 4.6 and 4.7. Tilting is here from [001] to [012] about the (200) diffraction vector. The beam direction in (*a*) is close to [001], in (*b*) is close to [013] and in (*c*) is close to [012]. The major axis of the loop turns clockwise, which is consistent with a circle on a $(1\bar{1}1)$ plane.

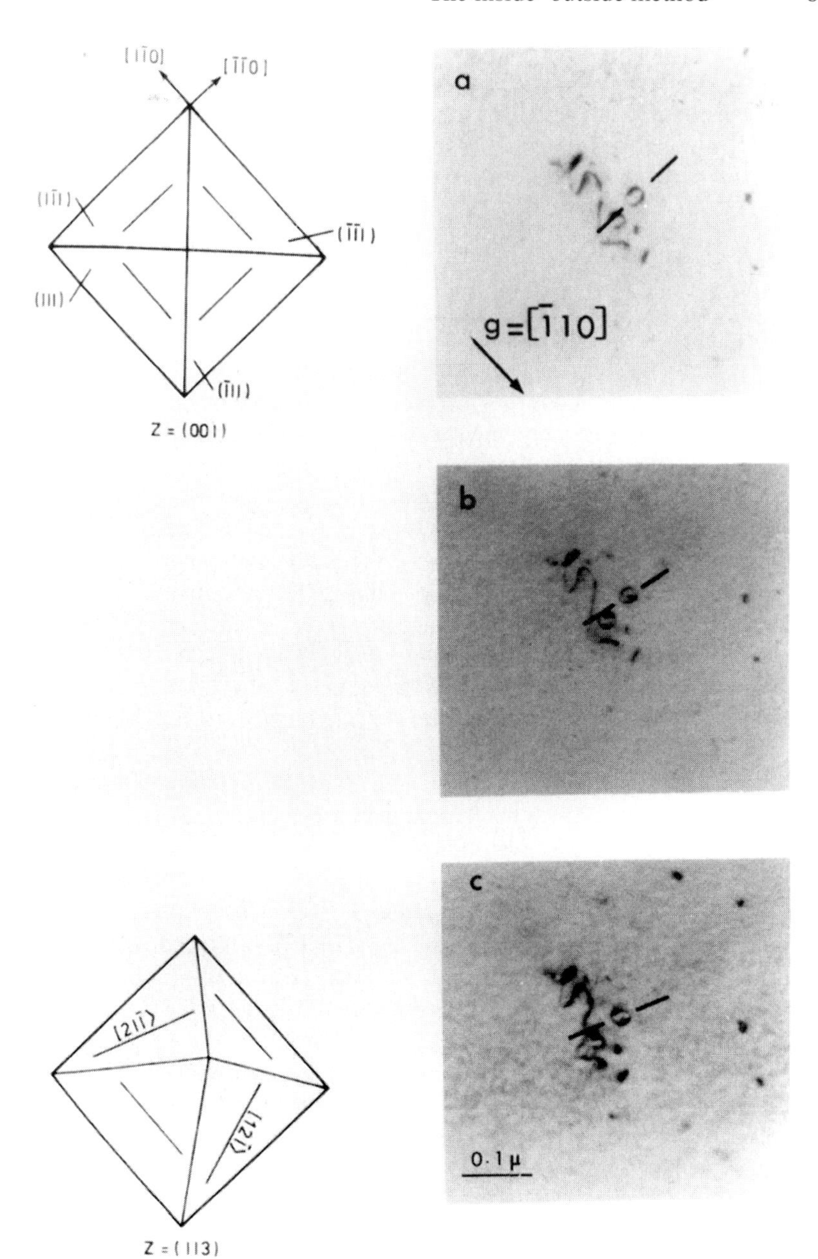

Figure 4.9. Further tilting experiment to confirm the habit-plane of the loop shown in figures 4.6–4.8. Tilting is here about the orthogonal $(\bar{1}10)$ diffraction vector. The beam directions for (*a*)–(*c*) are close to [001], [115] and [113] respectively. The major axis rotates clockwise, again consistent with a circle on a $(1\bar{1}1)$ plane.

of tilt as is done here. The technique allows a qualitative assessment of the loop normal $\boldsymbol{n}$. For both of the directions of tilting shown the change in shape of the loop is consistent with a circular loop lying within about 10° of the $(1\bar{1}1)$ plane. The loop is therefore pure-edge, or close to pure-edge, in character, and so the assumption in the previous analysis is confirmed to be correct.

4.2 The black–white stereo technique

Very small loops do not image as double arcs and rarely show well-defined inside–outside contrast behaviour. It is therefore difficult to apply the inside–outside method to dislocation loops smaller than about 10 nm in diameter, and it is certainly inapplicable to loops smaller than 5 nm. Dislocation loops in this size range lying close to either foil surface show so-called black–white contrast when imaged under dynamical two-beam conditions. In chapter 3 we noted that the sign of the scalar product $\boldsymbol{g} \cdot \boldsymbol{l}$ depends on the nature of the loop, and its depth in the foil, and this can be used as the basis for a nature determination. The depth dependence is shown for a vacancy loop in figure 3.2; the sign of $\boldsymbol{g} \cdot \boldsymbol{l}$ is opposite for an interstitial loop. Note that at the upper surface the direction of $\boldsymbol{l}$ reverses in the second layer L2 (which extends from about $0.35\xi_g$ to $0.7\xi_g$) compared with the first layer L1 (which extends from the surface to about $0.3\xi_g$). In the third layer L3 the contrast is similar to layer L1, although the black–white contrast in this layer may not be very distinct. The situation at the lower foil surface is similar, the only difference being that near this surface, bright-field and dark-field images are complementary rather than similar. The nature of a loop which shows black–white contrast can be determined if it can be placed unequivocally in one of the depth layers.

The determination of the depths of loops in the foil is usually made by stereo microscopy. Stereo pairs are usually recorded under weak-beam or kinematical bright-field conditions (although some researchers have used dynamical two-beam conditions). Two images are recorded at different projections using the same diffraction vector $\boldsymbol{g}$ and the same s_g. This is easily achieved using a modern goniometer stage by tilting the specimen about $\boldsymbol{g}$ along the Kikuchi band corresponding to $\boldsymbol{g}$. If possible, equal tilt angles on either side of the specimen normal are usually chosen. Careful measurements on the Kikuchi pattern allow the tilt angle to be measured to better than 1°. For a typical total angle of tilt of 10–25° and print magnification of $1\text{–}2 \times 10^5$, surface to defect parallaxes of a few millimetres are achieved. The relative depths of most individual loops can then be measured to an accuracy of about ±2 nm.

Various problems may arise with defect nature analysis by the stereo depth-measurement technique:

(a) The strength of the black–white contrast decreases with increasing defect depth. The black–white contrast is weaker for defects in the second and third depth layers L2 and L3 than in the first layer L1. Some defects in deeper layers do not show well-defined black–white contrast.

(b) In addition, loops lying in the transition regions between layers are expected to show weak, but perhaps rather complex contrast or appear as black dots (see figure 3.10). Sometimes the contrast of transition layer loops is so weak that these loops are effectively invisible. Loops lying further than about $1.3\xi_g$ from a surface do not show black–white contrast but are imaged as black dots.

(c) Whether a given defect shows well-developed black–white contrast may also be dependent on the exact deviation from the Bragg condition. Black–white images are obtained under dynamical two-beam conditions, with s_g small. However, in order to maximize the number of defects showing black–white contrast it is necessary to image the same area several times, varying the deviation parameter s_g by small amounts around $s_g \approx 0$.

(d) Errors in layer thickness. The extinction distance ξ_g may not be known accurately for the particular diffraction conditions chosen. Images are often not recorded at the exact $s_g = 0$ Bragg condition, although a correction may be made for this. In addition a many-beam value for ξ_g is required, rather than a two-beam value. Calculated extinction distances usually take no account of the presence of non-systematic reflections, which are often present in diffraction patterns.

(e) It is difficult to locate the surface accurately for depth measurements. A reference layer of gold islands is sometimes evaporated on the specimen to mark the surface used, but this may be deposited on a layer of oxide or of contamination, which may vary locally in thickness. This contamination layer can be more than 10 nm thick. The gold layer may also mask the contrast of small defects. If a reference gold layer is used, every effort should be made to deposit only on clean, unoxidized specimens.

(f) Errors in microscope magnification and in the tilt difference between members of the stereo pair, the necessity to correct for foil tilting effects, and subjective errors in the use of the stereo viewer all make it difficult to place all defects within depth layers with confidence.

Note that these problems are less acute if defects are located in the first depth layer L1. The technique is particularly difficult to apply to defects lying close to the centre of thicker foils.

For these reasons, a critical assessment of the black–white stereo technique was made by Fukushima *et al* (1997b). Their experiment was designed to investigate as completely as possible the application of the black–white stereo technique to the determination of the nature of the clusters produced by heavy-ion irradiation of copper.

Particular care was made to determine the loop depths as accurately as possible. Stereo micrographs were taken under weak-beam imaging conditions, which were found to give more consistent results than dynamical or bright-field kinematical imaging conditions. No surface markers were used, but the surface was defined by fitting of a plane just above (1 nm) a number of top defects. The plane parallel to the surface was calculated on the assumption that the depths of defects in the foil (their z-coordinates) were independent of their x–y-coordinates in the surface. Corrections were made for the fact that the stereo pairs were taken with the foil inclined to the beam. Average errors in the depth assignments of individual defects were estimated by measuring the depths of 130 defects twice, at different times and using different prints. The depth scale was confirmed by comparing the measured defect depth distributions with profiles generated by the modification by Bench *et al* (1991) and Frischherz *et al* (1993) to the Monte Carlo computer code TRIM (Biersack and Haggmark 1980). A number of other precautions were also taken in an attempt to optimize the method.

The main conclusions on the viability of the technique were as follows:

- In practice it is very difficult to apply the black–white contrast technique to whole cluster populations, that is, to loops of all sizes from the just-visible upwards, and to loops at all positions in the foil. This difficulty has largely been glossed over or ignored in previous investigations of defect nature using this technique, which have *not* examined whole cluster populations, but have concentrated (for example) on larger first-layer clusters.
- Attempts to change the layer structure by changing the operating reflection and microscope voltage to vary ξ_g met with limited success. Most clusters showed better black–white contrast or visibility in one or another of the imaging conditions, but only a few showed a change in the sign of $\boldsymbol{g} \cdot \boldsymbol{l}$, indicating a change in layer.
- Some clusters showed anomalous contrast, appearing apparently as vacancy in one condition but as interstitial in another. This may reflect a problem of placing individual defects within the layer structure. Even though the overall depth distributions seemed entirely reasonable, the errors on individual measurements may have been considerable, despite the precautions taken to minimize these errors. This may apply particularly to clusters of complex geometry and SFT, which may show complex or weak contrast under weak-beam imaging conditions, which were used for the stereo measurements.
- Some defects, which the analysis suggested to be interstitial in at least one diffraction condition, could be shown to be SFT or partially-dissociated Frank loops, and so were actually almost certainly vacancy in nature (see later). Clearly, conclusions based on the use of just one diffraction condition may be misleading.
- Finally, a large fraction of the defects (53%) eluded a nature identification. This is an important point, which has not been generally recognized. These clusters on average lie deeper in the foil than those which do show well-developed black–white contrast, but their size distribution is not very different from the population as a whole. It has always been known that a fraction of clusters lying in the dead zones between layers would show

black-dot or complex contrast in any given micrograph, but that such a large fraction would prove unanalysable under several different imaging conditions was unexpected. The problem may arise from the complicated geometry of the clusters in low stacking-fault energy fcc metals. It should be noted, however, that in *all* previous studies of neutron-irradiated materials typically only about half of the visible loops showed black–white contrast and so proved amenable to nature analysis. This is undoubtedly one of the biggest drawbacks of the technique.

In conclusion, then, if clusters do show well-developed black–white contrast, and their depth positions can be established with confidence, the black–white stereo technique works well. In practice, these criteria will usually be met by larger, first-layer clusters. Unfortunately, in many cases a large fraction of the clusters do not meet these criteria, either because they do not show well-developed black–white contrast or they lie deeper in the foil, and so their nature cannot be determined by this technique. The larger near-surface defects may well not be typical of smaller defects in the bulk, and so it is unwise to assume that the loops which cannot be analysed have the same nature as those which can. Some examples of the application of the technique follow.

4.2.1 Examples of use of black–white stereo analysis

Example 1: Conventional method—identification of deep-lying interstitial loops in copper and molybdenum produced by low-energy heavy-ion irradiations at low temperatures

Bullough *et al* (1991) carried out an experiment designed to investigate the role of replacement collision sequences (RCS) in cascade evolution. These authors irradiated single crystals of copper and molybdenum with high doses of low-energy heavy ions at 4.2 K and 78 K under well-controlled and characterized irradiation conditions. They then used the black–white stereo technique to characterize the vacancy or interstitial nature of the resulting loop populations after warm-up to room temperature. The investigators hoped to confirm earlier results of Hertel and co-workers (Hertel *et al* 1974, Hertel and Diehl 1976, Hertel 1979), who had reported evidence in support of the propagation of long RCS initiated in cascades. In copper irradiated with 5 keV Ar^+ ions at room temperature, Hertel *et al* found 'double-peak' interstitial loop depth distributions in specimens with [011] normals, and 'single-peak' distributions in specimens with [001] normals. These forms of distributions were interpreted as arising from the propagation of long RCS along close-packed $\langle 110 \rangle$ directions, with a mean range of about 18 nm and a small variance. It was argued that a double-peaked loop depth distribution arises at the [011] orientation because RCS propagating along the [011] direction normal to the foil surface deposit interstitials more deeply in the specimen than RCS propagating along the four shallowly-inclined $\langle 110 \rangle$ directions. At the [001] orientation however, all $\langle 110 \rangle$ directions not lying in the foil plane are equally inclined, so that RCS propagating along these directions would deposit interstitials at the same depth, and thus give a single-peaked depth distribution. Similar results in niobium were interpreted as arising from the propagation of long replacement sequences along $\langle 111 \rangle$ directions.

Bullough *et al* (1991) used fairly standard experimental techniques. Their stereo pairs were mostly taken under dark-field dynamical conditions, and they used gold islands to mark the position of the surface. They estimated that they could measure the depths of clearly imaged individual loops to an accuracy of about ± 2 nm. In retrospect this was perhaps optimistic. We now know that dynamical imaging conditions are not usually the best for stereo measurements, especially for very small defects.

Stereo data for an irradiated [110] copper specimen are shown in figure 4.10. The data illustrate some of the difficulties of the conventional technique previously alluded to. Several points should be noted. First, the marked position of the surface, estimated from the positions of the topmost defects, implies that the thickness of the condensate (upon which the gold islands were deposited) is about 13 nm. Second, the number of visible defects falls off at the layer boundaries, but not abruptly. The division between L2 and L3 is particularly fuzzy, probably as a consequence of difficulty in determining the depths of deeper defects. Black-dot defects do seem to occur more frequently close to the L1/L2 and L2/L3 boundaries for $\boldsymbol{g} = \bar{1}1\bar{1}$ but for $\boldsymbol{g} = 200$ are distributed throughout the layer structure. The nature of these black-dot defects cannot be determined explicitly from single micrographs. It was assumed somewhat arbitrarily by Bullough *et al* that defects not showing a clearly-defined $\boldsymbol{l}$-vector direction had the same nature as the majority of defects in the same layer, unless they lay near layer boundaries, in which case they were classified as 'not analysed'. Within the centres of both layers L1 and L2 there is a preponderance of images with $\boldsymbol{g} \cdot \boldsymbol{l} < 0$ for both reflections, although the situation for L3 is far less clear.

Despite these difficulties, Bullough *et al* (1991) were able to draw useful conclusions from this experiment, due largely to the fact that the vacancy and interstitial loop distributions were well separated in depth. Results for 10 keV ion irradiations of copper at 4.2 K are shown in figure 4.11. For all cases shown in this figure, a population of near-surface vacancy loops was found. Similar near-surface vacancy loop populations were seen for most irradiation conditions with ion energies ≥ 5 keV. These vacancy loops were considered to have formed by a process of cascade collapse. At ion doses greater than 5×10^{16} m^{-2}, populations of deeper-lying interstitial loops were identified. In molybdenum, only interstitial loops were visible, lying at a depth of about 10 nm below the surface. It was speculated that in this case, near-surface vacancy loops had been lost by glide to the surface. For both copper and molybdenum, the interstitial loop depth distributions depended sensitively on the incident ion mass, ion energy and specimen orientation, and were often rather broad. At ion doses $\geq 10^{17}$ m^{-2}, only a very small fraction ($\leq 2\%$) of the point-defects created at low temperature were retained in visible loops after warm-up to room temperature.

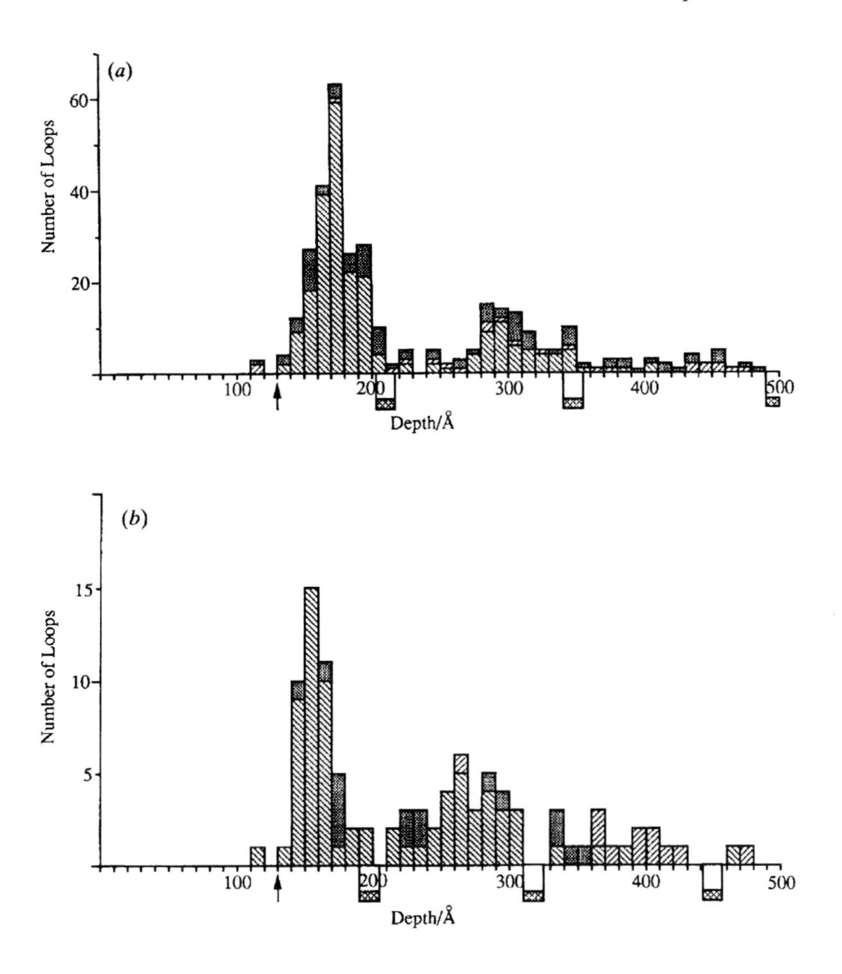

Figure 4.10. Stereo data from the experiment of Bullough *et al* (1991). The graphs show measurements of the sign of $\boldsymbol{g}\cdot\boldsymbol{l}$ for a [110] specimen of copper irradiated with 10 keV Cu^{+} ions at 4.2 K to a dose of 10^{18} ion m^{-2}, and imaged after warm-up to room temperature using (*a*) $\boldsymbol{g} = 200$, and (*b*) $\boldsymbol{g} = \bar{1}1\bar{1}$. The approximate location of the surface is indicated by the arrow. The predicted black–white contrast reversal depths are also indicated. Key to loop contrast: \\\, $\boldsymbol{g}\cdot\boldsymbol{l} < 0$; ///, $\boldsymbol{g}\cdot\boldsymbol{l} > 0$; cross-hatching, black-dot contrast.

An analysis of these results led Bullough *et al* to conclude that replacement collision sequences were playing at most a minor role. The observation that both vacancy and interstitial loop populations develop implies that vacancy–interstitial separation in cascades is relatively efficient, and this could be due to the propagation of RCS. However, no double peaks in the loop depth distributions were found either in copper or in molybdenum for any foil orientation or

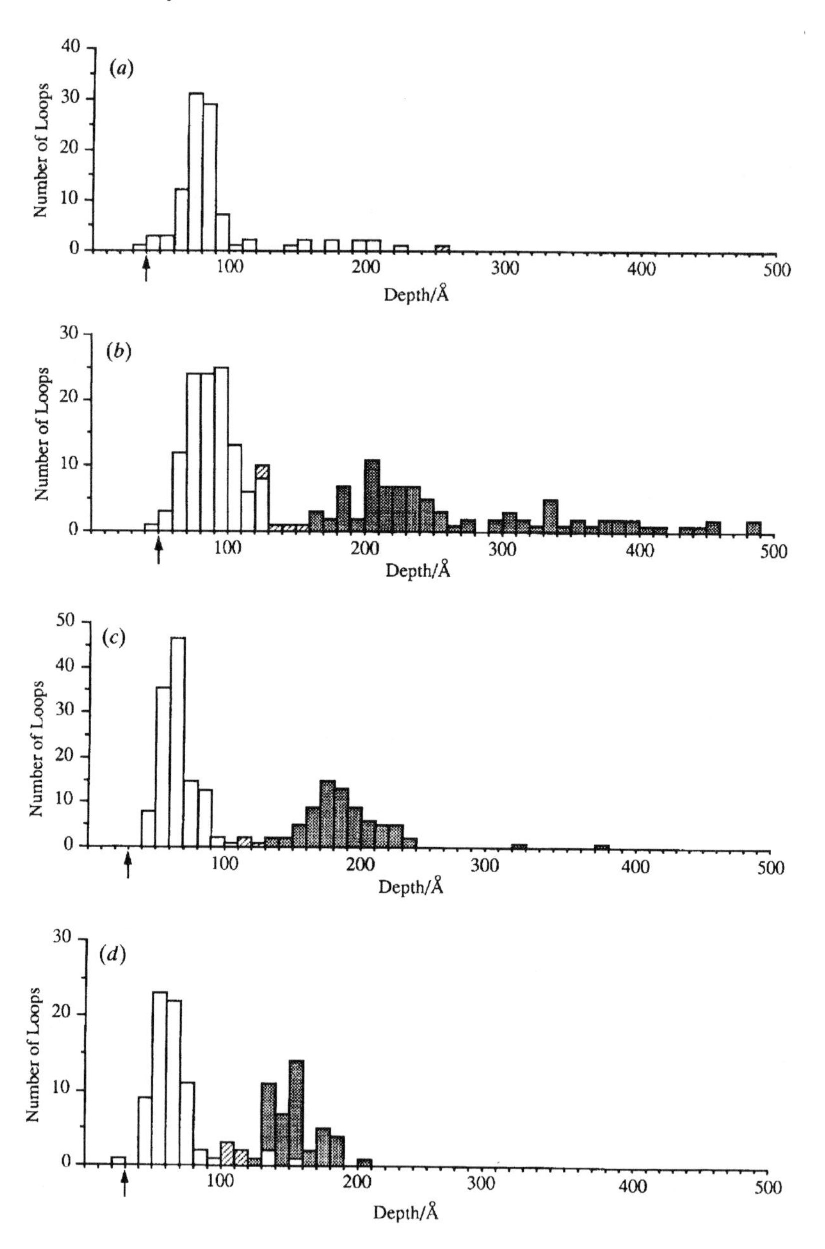

Figure 4.11. Loop depth distributions in copper after 10 keV heavy-ion irradiation at 4.2 K from Bullough *et al* (1991): (*a*) 5×10^{16} Cu^{+} m^{-2}, [011] surface; (*b*) 1×10^{18} Cu^{+} m^{-2}, [011] surface; (*c*) 7.5×10^{17} Xe^{+} ion m^{-2}, [011] surface; (*d*) 5×10^{17} Cu^{+} m^{-2}, [001] surface. □, vacancy loops; cross-hatching, interstitial loops; ///, uncharacterized loops; ↑, estimated position of surface.

irradiation condition. This implies that the range of replacement collision sequences is a few nanometres at most. The low point-defect retention efficiency is also consistent with a small average RCS range. Rather, it was considered that the retention of interstitials, and the form of the depth distributions of the interstitial loop populations, were determined by channelling of the incident ions at the surface in both copper and molybdenum. A possible explanation of the sharply-defined peaks found by Hertel and co-workers is heterogeneous nucleation of interstitial loops on argon atoms deposited by channelling.

Recent molecular dynamics simulations of cascades (e.g. Foreman *et al* 1992) have tended to support the conclusions of Bullough *et al* (1991). It is now generally accepted that replacement collision sequences play only a minor role in the evolution of high-energy displacement cascades.

Example 2: Modified method—the nature of cascade defects in heavy-ion irradiated copper

The experiment by Fukushima *et al* (1997b) discussed at the beginning of this section had the primary aim of determining the nature of clusters in copper, under heavy-ion irradiation. In most analyses of heavy-ion damage reported in the literature, the majority of defects were in the first layer (e.g. Wilson 1971, Häussermann 1972, Stathopoulos 1981). In these three (and in most similar) studies, the defects were reported to be of vacancy nature. Wilson (1971), however, notes that it was not possible to exclude the possibility that 'copper ion damage could contain a small proportion of interstitial loops' although none was identified. In recent years there has been increasing speculation, supported by some experimental observations, that interstitial loops can be produced in cascades. This possibility has been supported to some extent by computer simulations which show interstitial clustering within individual cascades, albeit that the clusters are of sub-microscopic size (De la Rubia and Guinan 1991, Foreman *et al* 1992, Bacon *et al* 1995).

In fact, the experiment confirmed that defects within about 10 nm of the foil surface were mostly or entirely vacancy in nature, as expected. Despite best efforts, the evidence for the presence of interstitial defects was equivocal. About 10% of the defects which could be analysed did appear to be more likely to be interstitial than vacancy on the criteria used. However, the evidence was relatively weak. Many of these possible interstitial loops lay deeper in the foil. It may be that this effect is real, that is, that interstitial loops are more likely to be produced well away from the surface, which acts as a strong interstitial sink. However, it seems equally likely that the result simply reflects the difficulty in placing some clusters correctly within the layer structure as suggested at the beginning of section 4.2. This difficulty arises for several reasons. First, errors in depth measurement for defects showing large parallax may be large, especially if the defect geometry is complex. The contrast of SFT, for example, can vary considerably between the two micrographs of the stereo pair, producing an illusory parallax difference. In other cases, the higher resolution micrographs showed the 'defect' to consist of two or more closely-spaced clusters, probably arising from separate subcascades. On the micrographs used for the stereo pairs these were often seen as a single image, but clearly the depth measurement in these cases cannot be trusted. Another problem was the possibility of confusing separate but closely-spaced defects on different micrographs despite the relatively low dose used. Finally, uncertainties in the positions of the layer structure itself increase with depth, making layer assignments of deeper defects particularly difficult.

This experiment does not answer the question of whether interstitial loops are present. The impossibility of showing that interstitial loops are *not* present is demonstrated by the fact that over half the total cluster population did not show sufficiently well-developed black–white contrast under any condition for a nature assignment to be made. Another attempt to identify interstitial loops is described in section 5.2.1.

Example 3: Use of black–white analysis to analyse cascade defects in YBCO without explicit measurements of defect depth

Heavy-ion irradiation has been proposed as a method of increasing the critical current density of the high-temperature superconductor $YBa_2Cu_3O_{7-\delta}$ (YBCO). The cascade-induced defects act as flux-pinning centres.

The defects produced by low-dose 50 and 85 keV Kr^+ and Xe^+ irradiation of YBCO have been shown by Frischherz *et al* (1993) to have primarily an amorphous-like structure. Under dark-field dynamical imaging conditions, these defects show black–white contrast with $\boldsymbol{l}$ parallel or antiparallel to $\boldsymbol{g}$. However, a direct determination of the nature of the strain field, that is inwardly-directed (vacancy-like) or outwardly-directed (interstitial-like), was not made because it proved difficult to measure defect depths by stereo microscopy. Instead, these investigators chose to model the defect depth using a modified TRIM computer code.

The basis of the technique can be appreciated from figure 4.12, which shows cascade defects produced (*a*) by irradiation with 50 keV Xe^+ ions, imaged in dark-field microscopy using $\boldsymbol{g} = 006$; and (*b*) by irradiation with 85 keV Xe^+ ions, imaged in dark-field microscopy using $\boldsymbol{g} = 200$. In both cases the same low dose (1.3×10^{15} ion m^{-2}) has been used, and both micrographs were taken at the exact Bragg condition ($s_g = 0$). In figure 4.12(*a*) the great majority of defects have black–white vector $\boldsymbol{l}$ parallel to $\boldsymbol{g}$. Three exceptions, which have $\boldsymbol{l}$ antiparallel to $\boldsymbol{g}$, are arrowed. In figure 4.12(*b*), the situation is almost reversed. The majority of defects (about 70%) have $\boldsymbol{l}$ antiparallel to $\boldsymbol{g}$, while the remainder have $\boldsymbol{l}$ parallel to $\boldsymbol{g}$. Clearly, these observations result from the defects in the two cases lying at different depths within the black–white layer structure.

The depth distributions in the two cases were calculated using a modified version (Bench *et al* 1991) of the Monte Carlo computer code TRIM-91 (Biersack

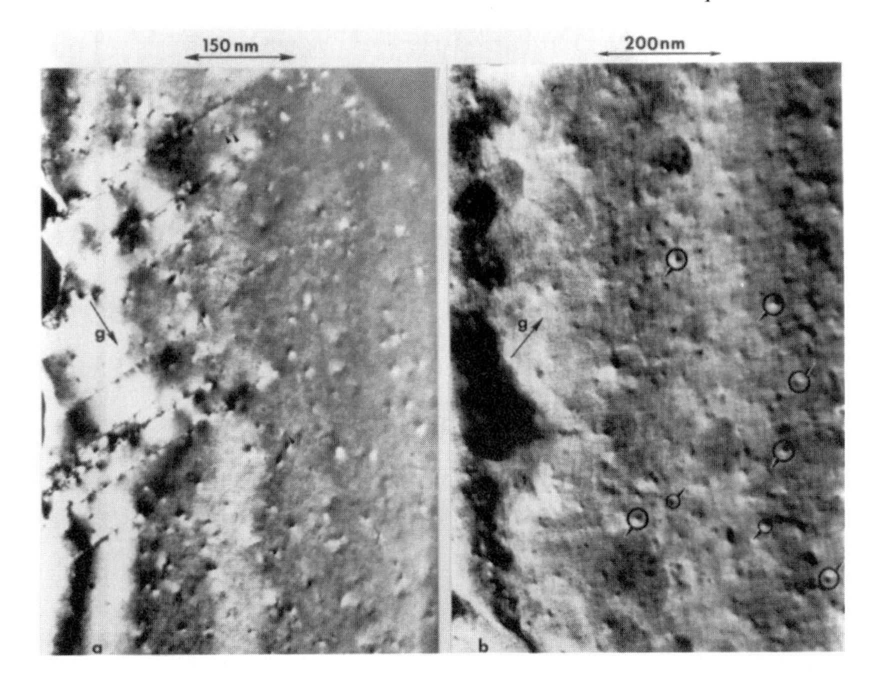

Figure 4.12. Dark-field dynamical images of YBCO specimens: (*a*) irradiated with 50 keV Xe^{+} ions and imaged using $\boldsymbol{g} = 006$; (*b*) irradiated with 85 keV Xe^{+} ions and imaged using $\boldsymbol{g} = 200$. In (*a*) a few of the defects with $\boldsymbol{l}$ antiparallel to $\boldsymbol{g}$ are arrowed. In (*b*) $\boldsymbol{l}$ is indicated for several defects. From Frischherz *et al* (1993).

and Haggmark 1980). The defect depths were characterized by the centre of the cascades as defined by Walker and Thompson (1978), i.e. by calculating the mean depth of the last displaced atoms in all secondary branches of the cascade. Then 10 000 ion cases were calculated to obtain sufficient statistics.

The results of the TRIM calculations are shown in figure 4.13. The layers for the (200) and (006) reflections are also indicated. The extinction distances $\xi_{200} = 40$ nm and $\xi_{006} = 53$ nm were calculated using the computer code EMS (Stadelmann 1987). Qualitatively it can be seen immediately that the distributions of parallel and antiparallel $\boldsymbol{l}$ are consistent with an inwardly-directed, vacancy-type strain field. For 50 keV ions and $\boldsymbol{g} = 006$, most defects lie in layer L1. For 85 keV ions and $\boldsymbol{g} = 200$, most defects lie in layer L2. This can be seen more quantitatively in table 4.1, where the distributions of defects predicted by TRIM are compared with those found experimentally, under the assumption of a vacancy-like strain field. The agreement is excellent, and it can be concluded with high confidence that the defects have strain fields of vacancy-type.

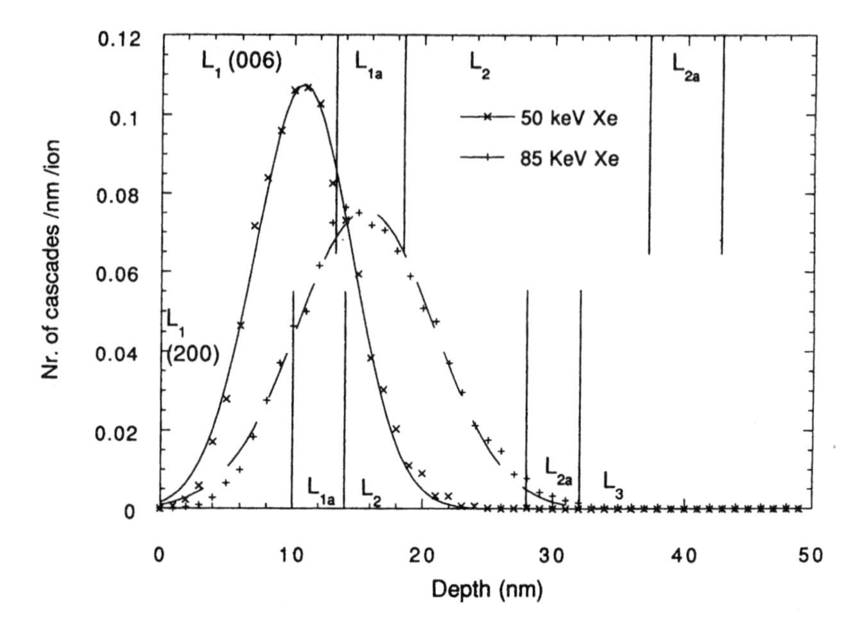

Figure 4.13. Depth distributions of cascades, as calculated by TRIM, for 50 keV and 85 keV Xe^+ ion irradiation. The layer structures for $g = 006$ and $g = 200$ are also indicated. From Frischherz *et al* (1993).

Table 4.1.

	50 keV Xe^+ $g = [006]$, TRIM calculated	50 keV Xe^+ $g = [006]$, measured	85 keV Xe^+ $g = [200]$, TRIM calculated	85 keV Xe^+ $g = [200]$, measured
L_1	70%, 95% p	>90% p	12%, 16% p	≈30% p
L_{1a}	26%		25%	
L_2	4%, 5% ap	<10% ap	64%, 84% ap	≈70% ap

4.3 The $2\frac{1}{2}D$ technique

Because of the experimental difficulties in applying the black–white stereo technique, and the impossibility of analysing a proportion of the defects when these lay deep in the foil, the radiation-damage community greeted warmly the introduction of an alternative method for nature characterization, the $2\frac{1}{2}D$ technique of Mitchell and Bell (1976).

This technique attempts to utilize differences in shifts of dark-field images of vacancy and interstitial clusters during defocusing of the objective lens.

The fundamentals of the $2\frac{1}{2}D$ technique have been described in the papers of Mitchell and Bell (1976), Michal and Sinclair (1979) and Sinclair *et al* (1981). If an image is taken using a diffracted beam $\boldsymbol{g}$ which makes an angle α with the optical axis of the microscope, and at a defocus D, then the image is shifted parallel to $\boldsymbol{g}$ by an amount by (Hirsch *et al* 1977)

$$y = (C_s\alpha^3 + D\alpha)\boldsymbol{g}/g = (C_s\lambda^3 g^2 + D\lambda)\boldsymbol{g}. \tag{4.1}$$

The defocus D is positive for overfocusing and negative for underfocusing. Here C_s is the spherical aberration coefficient, and $g = |\boldsymbol{g}| = \alpha/\lambda$, where λ is the wavelength of the electron beam. If two images are taken at different defocus values D_1 and D_2, then the relative shift $\Delta\boldsymbol{y} = (\boldsymbol{y}_1 - \boldsymbol{y}_2)$ of the images is given by

$$\Delta\boldsymbol{y} = (D_1 - D_2)\lambda\boldsymbol{g} = \Delta D\lambda\boldsymbol{g}. \tag{4.2}$$

A centred weak-beam dark-field image is obtained by tilting the incident beam such that the diffracted beam passes down the optic axis of the objective lens, and then placing an objective aperture centrally about the diffracted spot. Such images do not shift with a change of focus. If, however, an off-centre diffraction spot $\boldsymbol{g}'$ (due, for example, to a secondary phase or a crystalline defect) is also included in the objective aperture, a change of focus ΔD results in a shift of the image produced by $\boldsymbol{g}'$ relative to the image produced by $\boldsymbol{g}$ by an amount $\Delta\boldsymbol{y}'$. Then according to equation (4.2),

$$\Delta\boldsymbol{y}' = \Delta D\lambda(\boldsymbol{g}' - \boldsymbol{g}) = \Delta D\lambda\Delta\boldsymbol{g}'. \tag{4.3}$$

A vacancy cluster is expected to cause in its vicinity a local dilatational distortion whereas an interstitial cluster would cause a local compressional distortion. We may regard vacancy or interstitial clusters as causing a small local change in the diffraction vector $\boldsymbol{g}$: at vacancy clusters $\boldsymbol{g}_v$ is slightly shorter and at interstitial clusters $\boldsymbol{g}_i$ is slightly longer than the $\boldsymbol{g}$ vector for the perfect lattice. Following equation (4.3), the image shifts for the vacancy clusters ($\Delta\boldsymbol{y}_v$) and that for the interstitial clusters ($\Delta\boldsymbol{y}_i$) are given respectively by

$$\Delta\boldsymbol{y}_v = \Delta D\lambda(\boldsymbol{g}_v - \boldsymbol{g}) = \Delta D\lambda\Delta\boldsymbol{g}_v \tag{4.4}$$

$$\Delta\boldsymbol{y}_i = \Delta D\lambda(\boldsymbol{g}_i - \boldsymbol{g}) = \Delta D\lambda\Delta\boldsymbol{g}_i. \tag{4.5}$$

Consequently, the image shifts caused by overfocusing are in opposite senses for interstitial clusters and vacancy clusters, the former being parallel to $\boldsymbol{g}$ and the latter antiparallel if $\Delta D > 0$. This allows one, in principle, to identify the nature of clusters by analysing the direction of the shift of weak-beam images caused by a change of objective focus. If a pair of micrographs taken at different focus settings is viewed using a stereo viewer, the parallax produced by the image shifts results in defect clusters appearing at different apparent 'depths', depending on their nature and the sense of defocus.

The $2\frac{1}{2}D$ technique has been shown to work well for misfitting precipitates in determining whether the strain-field is compressional or dilatational (Mitchell and Bell 1976). However, the use of the $2\frac{1}{2}D$ technique for determining the nature of small loops and SFT has been criticized on a number of grounds. There are problems in principle: calculations by Grüschel (1981) (reported by Jenkins 1994) have shown that (1) the magnitude of the image shift with defocus may depend on the loop size, and its orientation in the foil as well as the sign and magnitude of its Burgers vector $\boldsymbol{b}$; and (2) image shifts may have a component perpendicular as well as parallel to $\boldsymbol{g}$. Piskunov (1985) has also commented that defocusing causes amplitude-phase effects which give rise to image shifts of the same order as the geometrical image shifts. There are also problems in practice: (1) it is difficult, if not impossible, to determine the zero-parallax plane, since even surface markers may shift with defocus; (2) the magnitude of the image shift may be small; and (3) faulted defects such as Frank loops or SFT give rise to reciprocal lattice spikes which will produce image shifts with defocus.

It is not clear whether these problems are of a magnitude to make the technique effectively unsuitable for analysing small clusters in all cases. However, Fukushima *et al* (1997a) have shown that the technique is not suitable for determining the nature of small clusters in ion-irradiated silver and copper. Since many if not all of the clusters in this case were expected to be faulted, problems associated with reciprocal lattice spikes lying perpendicular to the stacking fault planes of faulted loops were of particular concern. These authors therefore developed a modification of the $2\frac{1}{2}D$ technique which sought to recognize and avoid any influence of inclined reciprocal lattice spikes. However, this was not successful, and it was concluded that $2\frac{1}{2}D$ analyses of small faulted clusters was unreliable. The technique is also known to fail for small SFT (Sigle *et al* 1988).

At the time of writing, the question of whether the $2\frac{1}{2}D$ technique can be used to analyse successfully small *unfaulted* clusters is unresolved. Since the benefits would be large if the technique does work in this case, this would be worth pursuing.

4.4 Determining the nature of stacking-fault tetrahedra

The nature of SFT is of considerable interest because there is no reason in principle why interstitial SFT should not exist, although this may be unlikely since extrinsic stacking-fault energies are generally higher than intrinsic. In principle the black–white stereo technique can be applied to determine the nature of SFT (Saldin *et al* 1979), although as far as we are aware this has not been done. However, several other techniques have been used to determine the nature of SFT.

4.4.1 The method of Kojima *et al* (1989)

Kojima *et al* (1989) reported a relatively straightforward diffraction-contrast method which relies on the asymmetry of stacking-fault contrast under dark-field conditions when the deviation parameter $|s_g| \neq 0$. The method is illustrated in figure 4.14 which is taken from their paper. SFT are viewed in dark-field images under semi weak-beam conditions (typically $(\boldsymbol{g}, 2\boldsymbol{g})$ although the exact value of s_g is not important) along a $\langle 112 \rangle$ direction using a reflection of type 220. The two edge-on faults are out of contrast since they have $\boldsymbol{g} \cdot \boldsymbol{R} = 0$. The two inclined faults are visible as side-by-side triangles. The images of these two faults, however, are asymmetric: one is much brighter than the other, depending on the sign of $s_g(\boldsymbol{g} \cdot \boldsymbol{R})$. The fault displacement vector $\boldsymbol{R}$ is of different sign for extrinsic and intrinsic faults, as shown in the schematic diagram of figure 4.14(*e*). This image asymmetry allows the fault nature to be established.

Kojima *et al* (1989) found that the SFT in various materials (Au, Cu, Ag, Ni) produced by electron, neutron and ion irradiations were invariably vacancy in nature.

4.4.2 The method of Coene *et al* (1985)

Coene *et al* (1985) have described a high-resolution structural-imaging technique for determining the nature of SFT. The method relies on differences in the images of vacancy- and interstitial-type SFT under certain high-resolution imaging conditions. High-resolution structural images of SFT are formed with the foil oriented at an [011] zone axis. The experimental images are then compared with computer simulations. A distinction between SFT of vacancy or interstitial type can be made by examining the shift of the rows of bright dots in the high-resolution images.

The method is illustrated in figure 4.15. Figure 4.15(*a*) shows an SFT in silicon produced after P^+ ion implantation to a dose of 10^{20} ion m^{-2} followed by annealing at 700 °C for 3 hr. This high-resolution image was obtained near Schertzer defocus with the foil oriented at [011]. The SFT is visible as a V-shaped open triangle consisting of the $(1\bar{1}1)$ and $(11\bar{1})$ planes. Prominent bright dots show up along these planes. If the tetrahedron is viewed under grazing incidence in the $[0\bar{1}1]$ direction, that is, along the base of the open triangle, the rows of bright dots show a lateral shift away from the top of the open triangle. This shift is a maximum near the apex of the triangle and decreases to zero along the base. Image simulations for conditions corresponding to the experimental image are shown in figure 4.15(*b*) for an SFT of interstitial type, and figure 4.15(*c*) for an SFT of vacancy type. It can be seen that the observed image shifts are consistent with a vacancy nature. There is a good match between figure 4.15(*a*) and (*c*). An interstitial SFT would produce shifts in the opposite sense. Further image simulations for other defocus values are shown in the original paper, and show similar but sometimes less obvious image shifts.

The rather subtle contrast changes and their sensitivity to the exact defocus value make this technique more difficult to apply in practice than the technique of Kojima *et al*. However, Sigle *et al* (1988) have used the technique to investigate the nature of SFT near dislocations in electron-irradiated silver, and these were

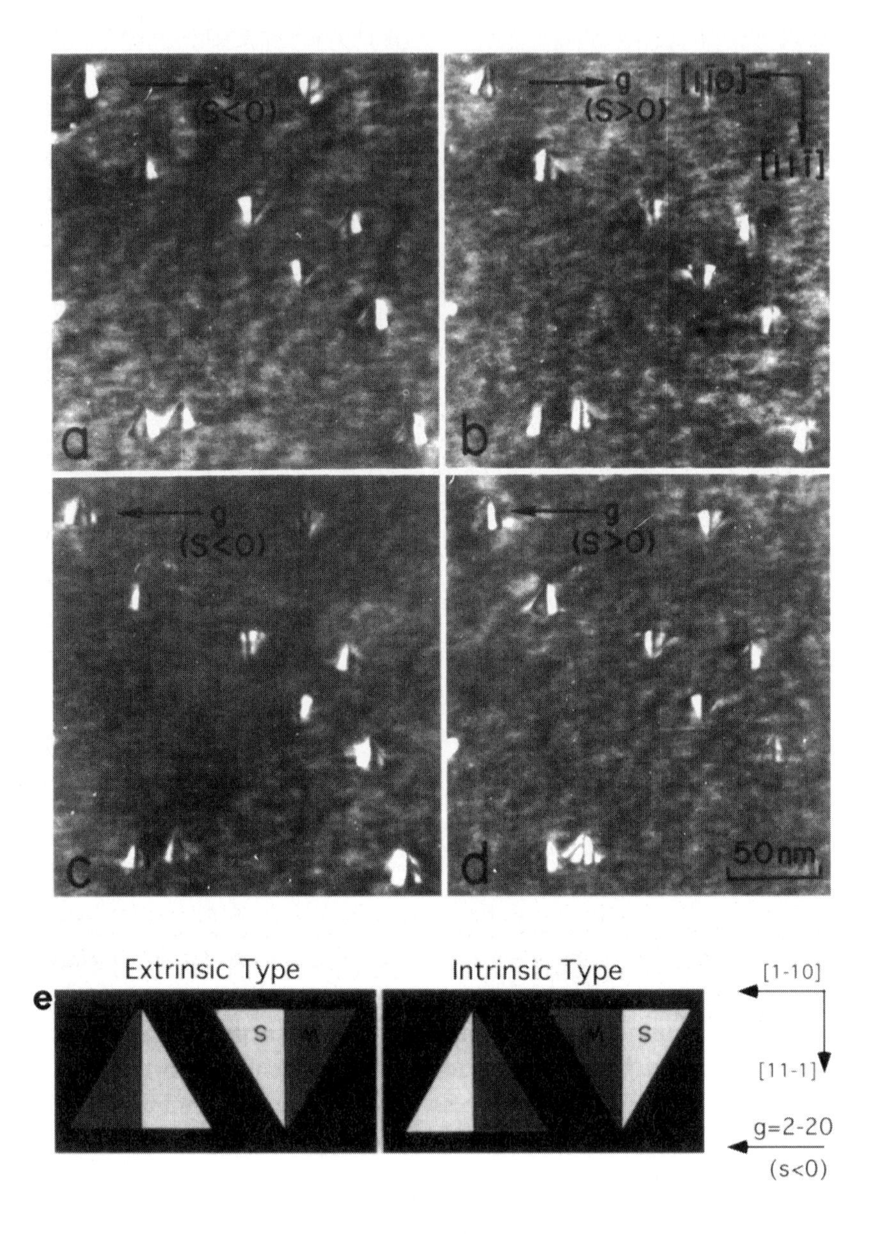

Figure 4.14. (*a*)–(*d*) 220 dark-field images of SFT in quenched gold, with different combinations of diffraction vector ***g*** and sign of deviation parameter s_g. (*e*) Schematic diagram showing the expected dark-field images for SFT of intrinsic (vacancy) and extrinsic (interstitial) type. The SFT in the micrographs of (*a*)–(*d*) are clearly of vacancy type. Figure modified from Kojima *et al* (1989).

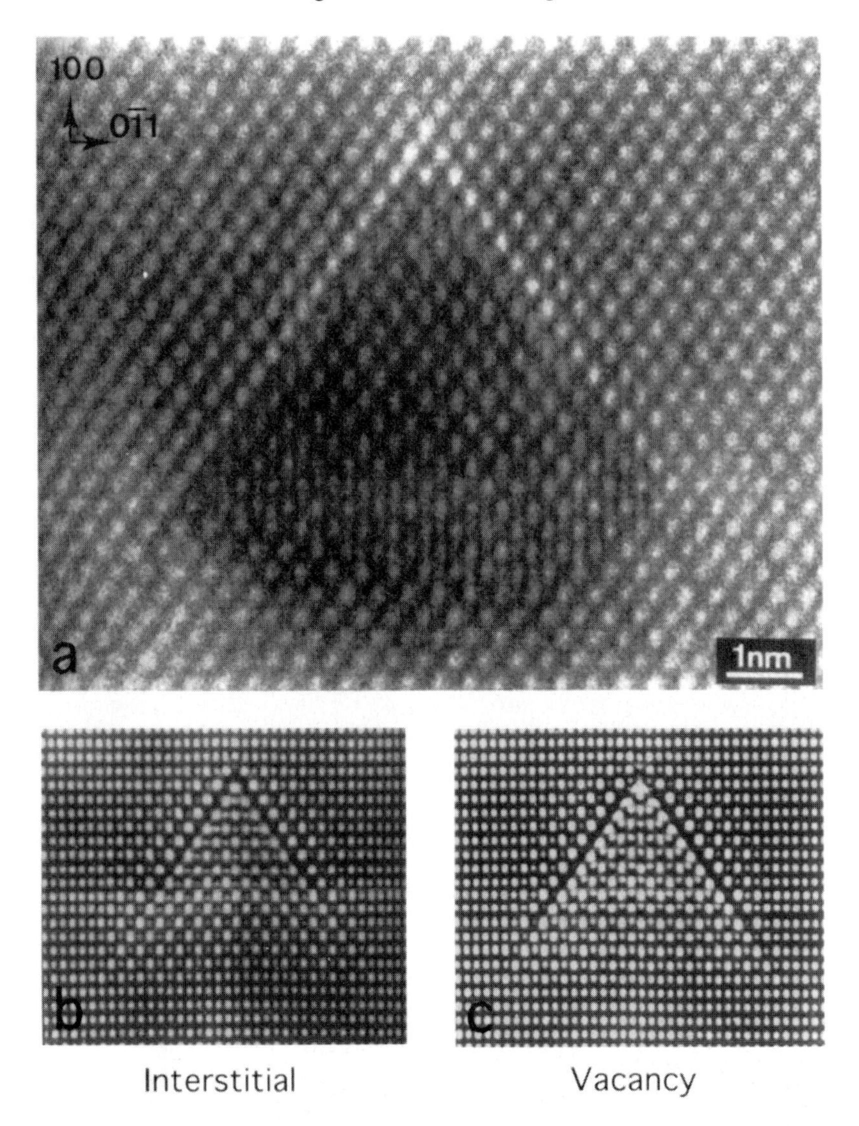

Figure 4.15. (*a*) A high-resolution structural image an SFT in P^+ ion-implanted and annealed (011) silicon foil. (*b*) Image simulations for an interstitial-type SFT, for an objective defocus of $\Delta f = -70$ nm. (*c*) As (*b*), but for a vacancy-type SFT. Comparison of the experimental and computer-simulated images (see text) shows that the SFT in (*a*) is vacancy in nature. From Coene *et al* (1985).

also found to be vacancy in nature. This result is consistent with the conclusions from an indirect technique, which is discussed in the next section.

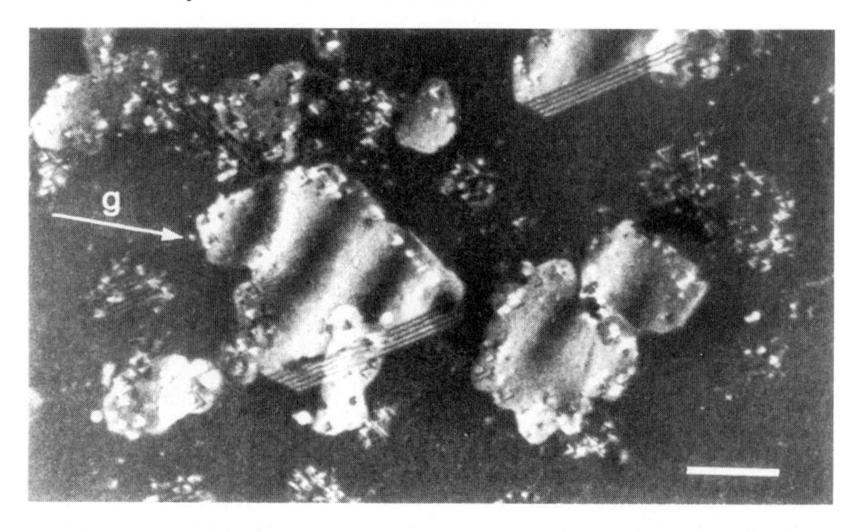

Figure 4.16. Formation of SFT at the inner peripheries of large pre-existing Frank loops in silver under 1 MeV electron irradiation. The scale marker is 50 nm. From Jenkins *et al* (1987).

In summary, there now seems to be a general consensus that all SFT formed under irradiation are vacancy in nature.

4.5 Indirect techniques for nature determination

Sometimes it is possible to infer the nature of small clusters indirectly using plausible arguments about the probability of nucleation and/or growth. In some cases, such indirect experiments may be subject to more than one interpretation, and so conclusions should always be treated with caution. Some examples follow.

(a) *Clusters which nucleate in regions of compressional strain* are more probably vacancy in nature, while clusters which nucleate in regions of dilatation are more probably interstitial. Figure 4.16 shows an *in situ* HVEM experiment to determine the nature of small SFT which form near dislocations in silver under room-temperature electron irradiation (Jenkins *et al* 1987). First, large interstitial Frank loops were produced by an elevated temperature 1 MeV electron irradiation. The loop nature was confirmed by the inside–outside technique. Then these loops were irradiated at room temperature, again with 1 MeV electrons. Small clusters, many of which can be recognized as SFT, formed inside the large loops, that is in regions of compressional strain. They are therefore more likely to be vacancy in nature.

(b) Arguments can also be made concerning clusters *which nucleate and grow near other sinks*. Thus loops which form in very thin regions of foil under

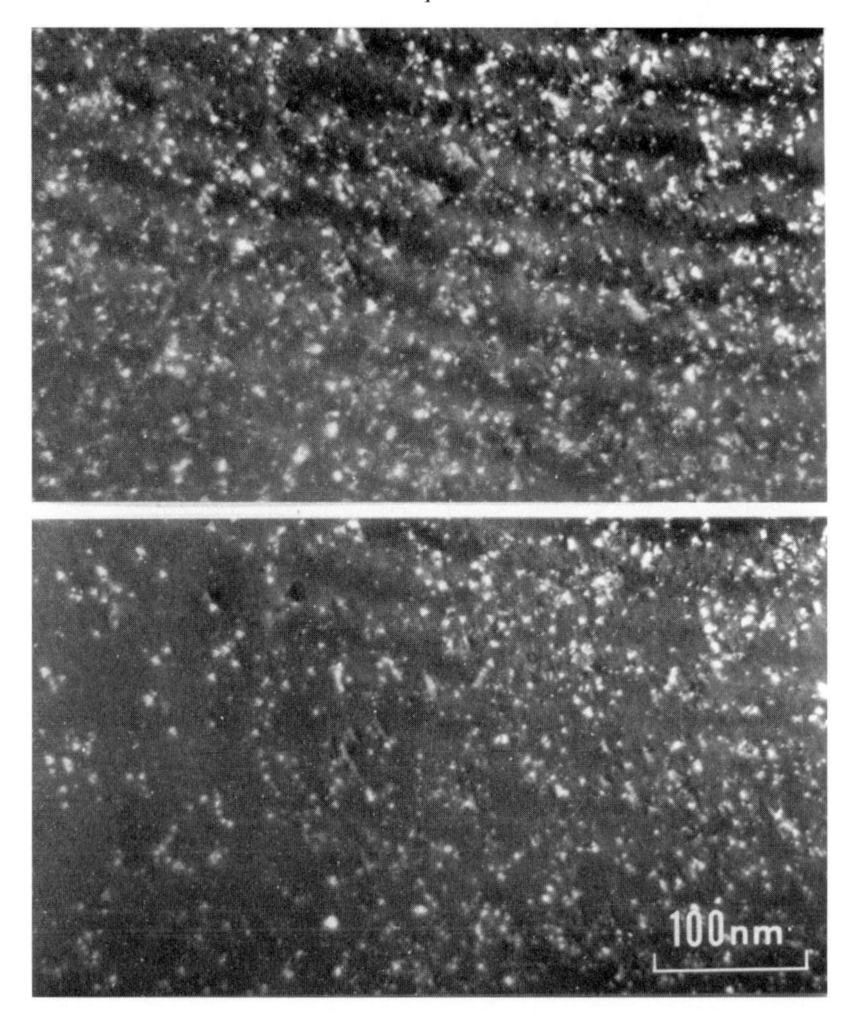

Figure 4.17. The top micrograph shows radiation damage in silver produced by 100 keV Cu^+ ion irradiation at room temperature. The lower micrograph shows the same area of foil, where the left-hand side only has been irradiated with 1 MeV electrons. In this area many of the clusters have shrunk or disappeared, showing they are most likely vacancy in nature. Micrographs from Kiritani *et al* (1994).

cascade-producing irradiation, under conditions where the foil surface is a dominant sink for the mobile interstitials, are more probably vacancy in nature. An example is described by Kiritani (1992) where in thin-foil irradiations of Ni with 14 MeV fusion neutrons at 473 K the clusters which are seen in the thinnest regions of foil are claimed to be vacancy in nature, whilst in thicker regions it was reported that vacancy and interstitial

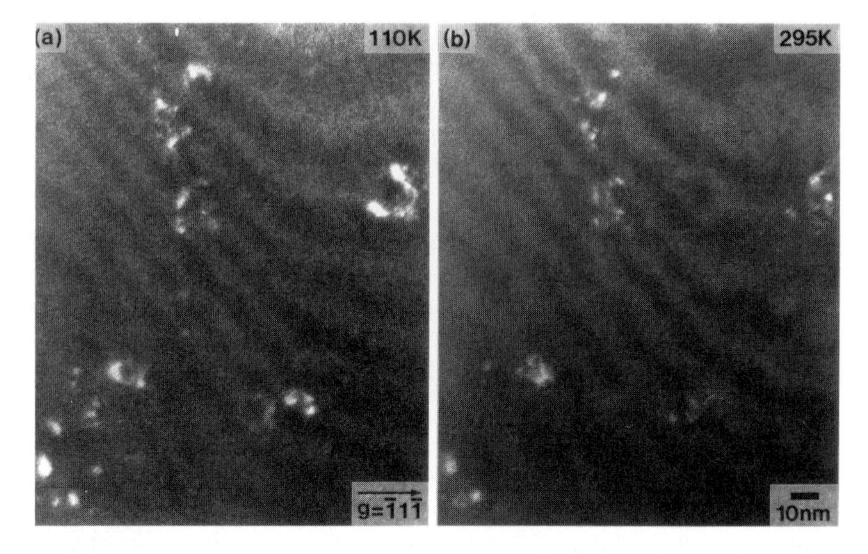

Figure 4.18. Defect clusters in silver irradiated with 14 MeV neutrons at 20 K to a dose of 9.6×10^{19} nm^{-2}: (*a*) imaged at 110 K after cryotransfer; (*b*) the same area after annealing at 295 K for 1 day. In both cases $\boldsymbol{g} = \bar{1}1\bar{1}$ with $s_g = 0.11$ nm^{-1}. Note the disappearance of a significant fraction of the clusters. Micrograph from Fukushima *et al* (1994).

clusters co-exist. The ground for believing this analysis is that the point-defect production rate under the irradiation conditions used was much lower than that required to form interstitial clusters in very thin foil by the agglomeration of freely-migrating interstitials. However, vacancy clusters may be formed directly by cascade collapse

(c) Cluster natures can be inferred from subsequent *in situ HVEM electron irradiation*. Pre-existing interstitial loops will tend to grow, and vacancy loops to shrink, provided that they do not lie too close to other sinks. The effect arises because the dislocation 'bias' for interstitials leads to a net interstitial flux to dislocation loops of both natures. An example is shown in figure 4.17 which is taken from the work of Kiritani *et al* (1994). The top micrograph in this figure shows small clusters in Ag produced by 100 keV Cu^+ ion irradiation at 300 K to a dose of 2.3×10^{16} ion m^{-2}. In the lower micrograph the left-hand side has been irradiated with 1 MeV electrons to a dose of 1.6×10^{24} m^{-2}. A large proportion of the clusters have shrunk or disappeared, implying that these are vacancy in nature. About 10% of the clusters appear to have grown under electron irradiation and these were considered by the authors to be interstitial in nature. This conclusion is perhaps less well established. Image sizes in weak-beam microscopy are very sensitive to the exact diffraction conditions, and so apparent small size increases may not be real (see section 5.2). In addition, clusters such as

SFT which are likely to be present have only a weak long-range strain field, and therefore a small bias. Such clusters may either grow or shrink due to stochastic variations in the arrival of interstitials and vacancies.

(d) The nature of clusters produced in silver and gold by low-temperature fission- and fusion-neutron irradiations of thin foils has been inferred from *in situ annealing experiments* in cryotransfer TEM (Fukushima *et al* 1989, 1994). Foils irradiated at 6 K or 20 K respectively were transferred at about 110 K to the cold stage of a conventional TEM. On annealing through stage III some clusters shrank and disappeared (figure 4.18). Since stage III corresponds to long-range vacancy migration, the disappearance was put down to recombination with vacancies. These clusters, more than 50% of the total, were therefore thought to be interstitial in nature.

Based on our more recent work (Jenkins *et al* 1999a), it is clear that in experiments of this type it is important to exclude the possibility of image artefacts. Weak-beam images of small defects may show apparent size changes or invisibilities with small changes in diffraction conditions (see figures 5.2 and 5.3). The problem can be circumvented by taking several images with small changes in deviation parameter. This is discussed in chapter 5.

Chapter 5

Analysis of small centres of strain: counting and sizing small clusters

This chapter is concerned with techniques for making quantitative measurements of the number densities, sizes and size distributions of small dislocation loops and SFT. Voids and bubbles will be considered separately in chapter 6.

5.1 Determination of loop and SFT number densities

Accurate determinations of loop and SFT number densities are difficult for a number of reasons. These difficulties will now be enumerated and some methods of alleviating them discussed.

5.1.1 There is a finite resolution limit—some loops are not seen because they are too small

The resolution limit will depend on the foil quality, the imaging conditions, and possibly the type of defect. In the best foils, loops of diameter $d \geq 1$ nm are visible under appropriate diffraction conditions—weak-beam conditions generally seem best, although it is necessary to record several images with varying s_g in order to image as high a fraction of loops as possible. This will be discussed in more detail in section 5.2. SFT with edge lengths ≥ 1 nm are also visible and identifiable in good-quality foils under appropriate weak-beam or kinematical diffraction conditions. If the foil quality is poor, the resolution limit may be much worse. The problem is fundamental and so cannot be avoided altogether, although obviously considerable attention should be paid to optimize the quality of the foil. Image size distributions may be helpful in indicating if the problem in a particular case is serious. If the distribution peaks at small image sizes close to the estimated resolution limit, this may indicate the presence of a large unresolved loop population.

5.1.2 In any one micrograph, only a proportion of the resolvable loops will be seen

Some loops will be out-of-contrast (i.e. $\boldsymbol{g} \cdot \boldsymbol{b} = 0$) and so will show only weak contrast which may be difficult to discern from background (e.g. in dynamical conditions, 'butterfly' contrast). Others may show weak contrast because they lie at unfavourable depths, for example at the transition zone between depth layers. The problem will be particularly severe for defects of size close to the resolution limit.

This problem can be alleviated by imaging the same area under different diffraction conditions, using several different diffraction vectors and deviation parameters s_g. This is tedious, and so often counting is made from a single micrograph, and a correction made for loops which are out-of-contrast in this condition. Difficulty may arise if loops with $\boldsymbol{g} \cdot \boldsymbol{b} = 0$ are still visible because $\boldsymbol{g} \cdot \boldsymbol{b} \times \boldsymbol{u} \neq 0$, but in practice such contrast can often be recognized.

Particular diffraction vectors may be more suitable than others for seeing most of the loops. In the case shown in figure 2.2 of the analysis of loops in an (011) foil of copper using several diffraction vectors, all Frank loops ($\boldsymbol{b} = \frac{1}{3}\langle 111 \rangle$) should be in contrast ($\boldsymbol{g} \cdot \boldsymbol{b} \neq 0$) in figure 2.3(*c*) taken in $\boldsymbol{g} = 200$, and four of the six prismatic loop variants ($\boldsymbol{b} = \frac{1}{2}\langle 110 \rangle$) should also be in contrast. Only the two variants $\boldsymbol{b} = \pm\frac{1}{2}[011]$ and $\boldsymbol{b} = \pm\frac{1}{2}[01\bar{1}]$ have $\boldsymbol{g} \cdot \boldsymbol{b} = 0$, and so would be out-of-contrast. (In fact, in this case prismatic loops are not present, although that would not have been known when the experiment was carried out). The reflections $\boldsymbol{g} = 1\bar{1}1$ and $\bar{1}\bar{1}1$ also have $\boldsymbol{g} \cdot \boldsymbol{b} \neq 0$ for all of the Frank loops, but would image only three of the prismatic loop variants. The image taken in $\boldsymbol{g} = 0\bar{2}2$ is least suitable for number density measurements since it puts out of contrast two of the four Frank loop variants.

Even when $\boldsymbol{g} \cdot \boldsymbol{b} \neq 0$, small loops are not always visible. An experiment where s_g is varied from zero (dynamical two-beam conditions) to small positive values (bright-field kinematical conditions) is shown in figure 2.4. It can be seen that although $\boldsymbol{g}$ is the same for each micrograph, no one micrograph picks up all the loops. A comparison of several micrographs enables weak features to be identified as loops or not. The usual criterion is that if a feature appears with similar contrast in two or more micrographs, it is more likely to be a loop than an imaging artefact. It will be seen (section 5.2 and figure 5.3) that the situation is similar for weak-beam imaging, although weak-beam imaging is generally preferable for very small clusters.

Another possibility for maximizing the proportion of loops seen in a single micrograph is to use 'down-zone imaging' (section 2.1). Even in this case, however, some loops are not visible, probably because they lie at unfavourable depths. In practice, down-zone images seem to pick up fewer small clusters than the best weak-beam images (see figure 2.5).

5.1.3 Loops may be lost from the foil due to surface image forces

This happens, for example, in heavy-ion irradiations of bcc and hcp metals, where the vacancy loops produced by the collapse of displacement cascades unfault to glissile prismatic loops. Jäger and Wilkens (1975) have shown that the presence of a nearby surface may cause two effects. First, it may influence the unfaulting reaction. A given faulted loop of type $\boldsymbol{b} = \frac{1}{2}\langle 110\rangle$ can unfault to one of three different $\boldsymbol{b} = \frac{1}{2}\langle 111\rangle$ prismatic loop variants. The most likely variant to be produced is that with the largest component of $\boldsymbol{b}$ towards the surface. Second, if the elastic interaction energy of the loop with the surface is sufficiently high, surface image forces cause the prismatic loop to glide to the surface and be lost.

Daulton *et al* (2000) have recently studied the loss of loops to the surface during *in situ* ion irradiation of copper at elevated temperatures. Using video recording they found enhanced loop loss during irradiations, observed to occur within one video frame (0.03 s), but occasionally some motion was observed over a few seconds before loss. The loss fraction increased with irradiation temperature (measured up to 873 K). It is believed that both factors of the near surface and energy input from nearby irradiation events contribute to unfaulting and glide to the surface of these loops.

Loss of loops to the surface will generally be manifest if a loop geometry analysis is performed. In bulk material it would be expected that all possible loop variants occur in equal proportions. If this is not the case, surface effects should be suspected. It is possible to make a correction, but both effects listed here have to be taken into account.

5.1.4 Counting may be difficult if the loop number density is very high

This problem may often be alleviated by the use of weak-beam rather than strong-beam diffracting conditions. The narrower peak widths in weak-beam images make image overlap problems less severe.

An interesting alternative approach to this problem may be applicable if the loops are faulted, which is the case, for example, in austenitic steels after neutron irradiation in the temperature range from 300 to 550 °C. In this case it is possible to image the loops using fine-structure diffraction effects associated with the stacking faults.

The method is illustrated in figure 5.1, which is taken from the work of Brown (1976). Figure 5.1(*a*) shows a bright-field image of a neutron-irradiated M316 stainless steel containing a high number density of faulted Frank dislocation loops. Clearly this image is of little use for determining the number density or sizes of the loops and their distribution over different planes. Figure 5.1(*b*) shows a selected area diffraction pattern. The foil is oriented close to [100] and it is apparent that satellite spots are present around the four {200} reflections. The origin of these satellite spots are reciprocal lattice spikes associated with stacking faults on the inclined {111} planes, as shown

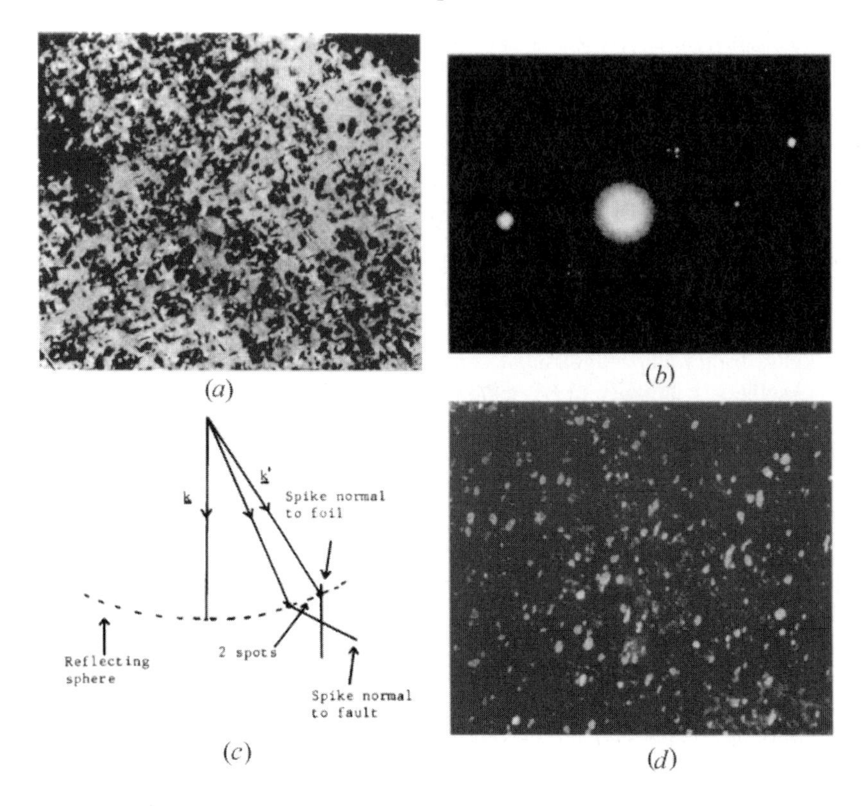

Figure 5.1. Faulted Frank loops in neutron-irradiated stainless steel: (*a*) bright-field image, with all four loop variants visible; (*b*) selected-area diffraction pattern, showing satellite spots around the 002 and 020 reflections; (*c*) schematic showing the origin of one satellite spot; (*d*) dark-field image obtained using one of the satellite spots, showing only one set of Frank loops. From Brown (1976).

schematically in figure 5.1(*c*). The satellites will be visible around a given {200} reflection only if the reflection is not strongly excited, so that its reciprocal lattice point is not intersected by the Ewald sphere. Five reciprocal lattice 'spikes' are associated with each reciprocal lattice point. One is in a direction normal to the thin foil, and the intersection of this spike with the Ewald sphere gives rise to the central {200} reflection. The other four spikes are normal to the {111} planes of the stacking faults, and it is the intersection of these inclined spikes with the Ewald sphere which gives rise to the satellite spots. Each satellite spot originates from the stacking-faults of Frank loops on a particular {111} plane. It is possible by careful positioning of a small objective aperture to form a dark-field image using one of the satellite reflections. The result of doing this is shown in figure 5.1(*d*). Now just one of the four sets of Frank loops is visible. Clearly this image is more suitable for counting and sizing this set of loops. By forming similar images using the other satellite reflections the other three sets of loops can be analysed in the same way.

5.1.5 Foil-thickness measurement will be necessary if volume (rather than area) number densities are needed

In much of the earlier work in the literature, the foil thickness was estimated rather crudely, for example, by counting the number of thickness fringes from the foil edge. If say a dark-field micrograph is recorded in the third bright thickness fringe under dynamical conditions, the foil thickness is about $1.5\xi_g$. This method is unlikely to be accurate to better than about $\pm 20\%$. It is still used frequently because of its simplicity. It is also possible to estimate the foil thickness using stereo microscopy if features such as stacking-faults, which intersect both surfaces, are present. This method is now rarely used. Rather, with the advent of modern analytical electron microscopes equipped with condenser-objective lenses, it is straightforward to measure the local foil thickness much more accurately by the use of a simple convergent-beam technique (Kelly *et al* 1975). This technique involves measuring the spacing of Kossel–Möllenstedt fringes in the dark-field convergent-beam disc with the foil set at a two-beam orientation. Experimental details are given by Williams and Carter (1996, chapter 21). In thicker foils, where several Kossel–Möllenstedt fringes can be measured, the foil thickness can easily be determined to a few percent, or better if many-beam effects are taken into account (Castro-Fernandez *et al* 1985).

In practice, it is sometimes necessary to use very thin foils, for example when using the disordered zone technique described in chapter 7. In this case only one or two fringes may be visible in the dark-field convergent-beam disc, but even so it is possible to estimate the foil thickness with reasonable accuracy. The way to do this is to record at the same time one or more convergent-beam patterns from thicker regions of foil. The plots of $(s_i/n_k)^2$ versus n_k^2 (see Williams and Carter (1996) for notation) for different foil thicknesses t are straight lines, with the *same* slope $(1/\xi_g)^2$ but different intercepts on the y-axis $(1/t^2)$. Even if the foil is so thin that only one point can be plotted, a straight line can be drawn through this point with the same slope as the plots for the thicker regions, and the intercept measured with fair accuracy.

5.1.6 For defect yields in irradiation experiments, it is also necessary to consider errors in dose measurement

Dosimetry in many experiments may have considerable uncertainties. The magnitude of these uncertainties is difficult to estimate, and is not strictly a concern of this book. However, in some irradiation experiments two possible approaches may be adopted to circumvent the difficulty. The first is to use *comparative* yields only. Thus a systematic series of experiments may be defined where, say, various materials are irradiated with heavy-ions simultaneously. The relative yields will clearly be subject to more confident interpretation than absolute yields obtained in the same experiment, or by other workers. If, however, *absolute* yields are required, it is possible to use a built-in dose monitor. In heavy-ion irradiations, specimens of Cu_3Au or Ni_3Al may serve this purpose. In these ordered alloys it is possible to image the 'disordered zones' created at cascade sites by the impact of each ion. The technique works well for low doses, say $<5 \times 10^{11}$ ion m^{-2}, where individual ion impacts are well separated. The conditions for doing this and some further details are given in chapter 7.

5.2 Determining loop and SFT sizes

Determining the sizes of *individual* small dislocation loops (<10 nm) is difficult. The relationship between the image size and the true loop size may depend on the foil thickness and the depth of the loop in the foil, neither of which is known accurately, as well as the exact diffraction conditions.

Dislocation loops lying close to the foil surface are often sized from dynamic two-beam micrographs, where they show black–white contrast. Katerbau (1976) has shown by image simulations that the length of the black–white interface of dynamical images gives a reasonable estimate of the true loop size. The *average* value of the ratio between image width and defect diameter is unity, although for individual defects this ratio may vary by up to 50%. It is therefore possible to generate loop-size *distributions*, at least for those loops which show black–white contrast. As emphasized in chapter 4, many of the smallest loops and those lying towards the centre of thicker foils do not show well-developed black–white contrast, and so will not be included in the distributions. Systematical errors may result, which are not always recognized.

When loops appear as black dots (e.g. under kinematical bright-field conditions, or when they lie between or outside the black–white layers) the maximum dimension of the black dot is often taken as the loop diameter. The justification for this is dubious. Indeed Wilkens *et al* (1973) have shown that the image sizes under different 'kinematical' conditions can vary considerably. However, an analysis of defects in the high-temperature superconductor $YBa_2Cu_3O_x$, produced by 50 keV Kr^+ ion irradiation suggests that, in some cases at least, this approach is not unreasonable. Storey *et al* (1996) measured the size distributions of clusters showing black–white contrast and black-dot contrast on the same micrograph, and found that the overall shapes of the two distributions were nearly identical, and the shifts in the peaks was <0.3 nm (figure 5.2).

For SFT, the situation with respect to sizing under bright-field kinematical conditions seems better. Although no specific calculations seem to have been reported, SFT of edge length ~2 nm can certainly be imaged under kinematical conditions as well-defined triangles or squares (depending on their projected shape). Examples are shown in figure 2.3. Since SFT are imaged mainly by

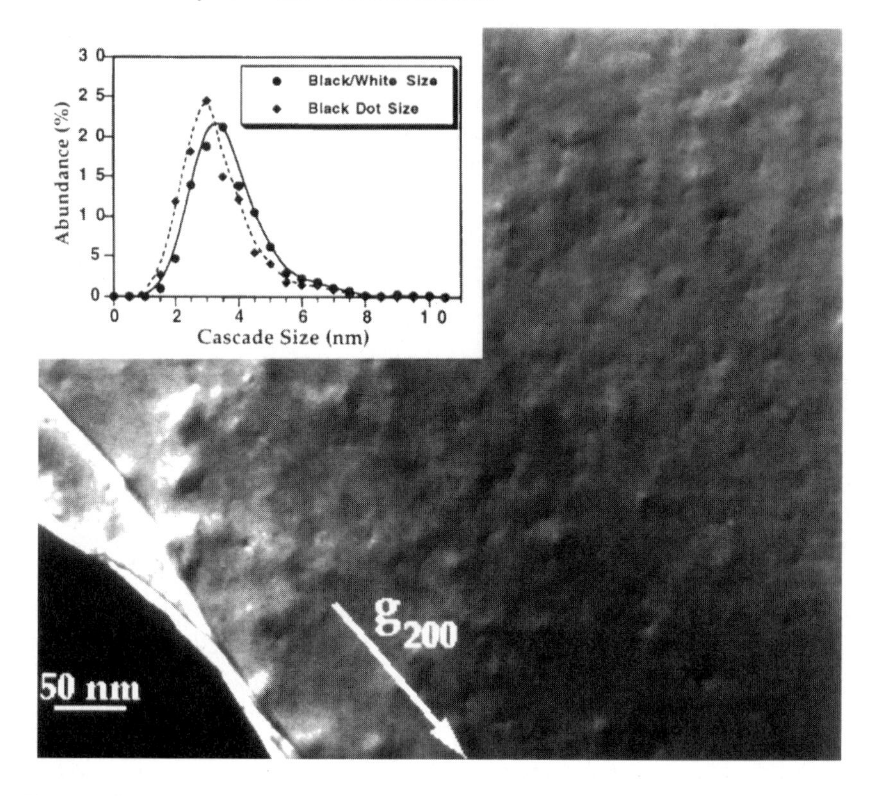

Figure 5.2. Size measurements (insert) of defect cascades created in YBCO by 50 keV Kr^+ ion irradiation. The distributions of separate measurements from defects with black–white contrast and those with black dot contrast are shown to be very similar. The complete experiment showed the average cascade defect size to be independent of oxygen concentration in YBCO, in contradiction to a theory of magnetic flux pinning in this high-temperature superconductor. From Storey *et al* (1996).

the stacking-faults on their faces, and not by the weak $\boldsymbol{b} = \frac{1}{6}\langle 110 \rangle$ stair-rod dislocations along their edges, it seems possible that such kinematical images do define accurately the extent of the tetrahedron. However, no image calculations or experiments have been made to confirm this.

In practice, very small loops, which often show very weak contrast or are completely invisible under two-beam dynamical or bright-field kinematical imaging conditions, frequently show much better visibility under weak-beam conditions. The sizes of loops larger than about 10 nm can often be determined from measurements made on single weak-beam micrographs, although great care is necessary in setting the diffraction conditions and interpreting the image. Jenkins *et al* (1973—see section 3.2.1, example 1, and figure 3.18) showed that the sizes of 10–20 nm Frank loops with $\boldsymbol{b} = \frac{1}{3}[111]$ in a (111) foil of silicon could be determined accurately by recording images in $\boldsymbol{g} = 2\bar{2}0$ with the loop flat-on. In this case $\boldsymbol{g} \cdot \boldsymbol{b} = 0$, but segments of the loop with $\boldsymbol{g} \cdot \boldsymbol{b} \times \boldsymbol{u} \neq 0$ show characteristic double peaks which delineate the position of the core of the dislocation line accurately. More generally for loops in this size range, the maximum image size usually seems to give a reasonably good indication of the true size of the defect, although this has not been explored systematically by image simulations. Experimentally, it is known that complications can arise if defects have complex geometry. For example, under some imaging conditions partially-dissociated Frank loops may show two or more image peaks, which could be mistaken for separate defects (see section 3.2.1, example 2). Taking images using several different diffraction vectors allows such effects to be recognized.

Weak-beam imaging must be used with caution for very small loops ($d <$ 10 nm), although this may be the only practicable alternative. We have recently made an experimental analysis of the conditions necessary for the successful use of the weak-beam technique for identifying and sizing small ($\leq$5 nm) point-defect clusters in ion-irradiated copper (Jenkins *et al* 1999a). The influence of several potentially important parameters was assessed. In general, the image sizes of small clusters were found to be most sensitive to the magnitude of s_g, with the image sizes of some individual defects changing by large amounts with changes in s_g as small as 0.025 nm^{-1}. Individual loops, clearly visible under one condition, may even be totally invisible under another. This can be seen in figure 5.3. The visibility of small defects was found to depend only weakly on the beam convergence, however, which is a considerable benefit since it is usual to record weak-beam images with the second condenser lens fairly well focused in order to reduce exposure times.

Jenkins *et al* (1999a) made a careful analysis of the ten defects marked on figure 5.3. In addition to the series of micrographs shown, three further series of micrographs were taken, for all combinations of the signs of both $\boldsymbol{g}$ and s_g. Each negative was printed to the same total magnification 5×10^5, taking care to maintain consistent print exposures. The prints were selected in random order, placed in turn on a digitizing tablet and examined through a 10× eyepiece. In general, the maximum extent of each image was measured, although care was taken always to measure the same defect in the same direction on each print. Each size measurement was repeated 10 times in order to improve the statistical accuracy. It was concluded that defects in copper could be reliably revealed by taking a series of 5–9 micrographs with a systematic variation of deviation parameter from 0.2–0.3 nm^{-1}. The maximum image size could be measured to an accuracy of about 0.5 nm for defects situated throughout a foil thickness of 60 nm. The maximum image size may represent the best estimate of the true defect size, although this has yet to be confirmed by contrast calculations. Examples of size measurements are shown in figure 5.4. The procedure is clearly very tedious, but necessarily so. Image sizes for the same defects varied considerably from one micrograph to another, and so automation of size measurements by computer would probably not work. The technique has been applied to the determination

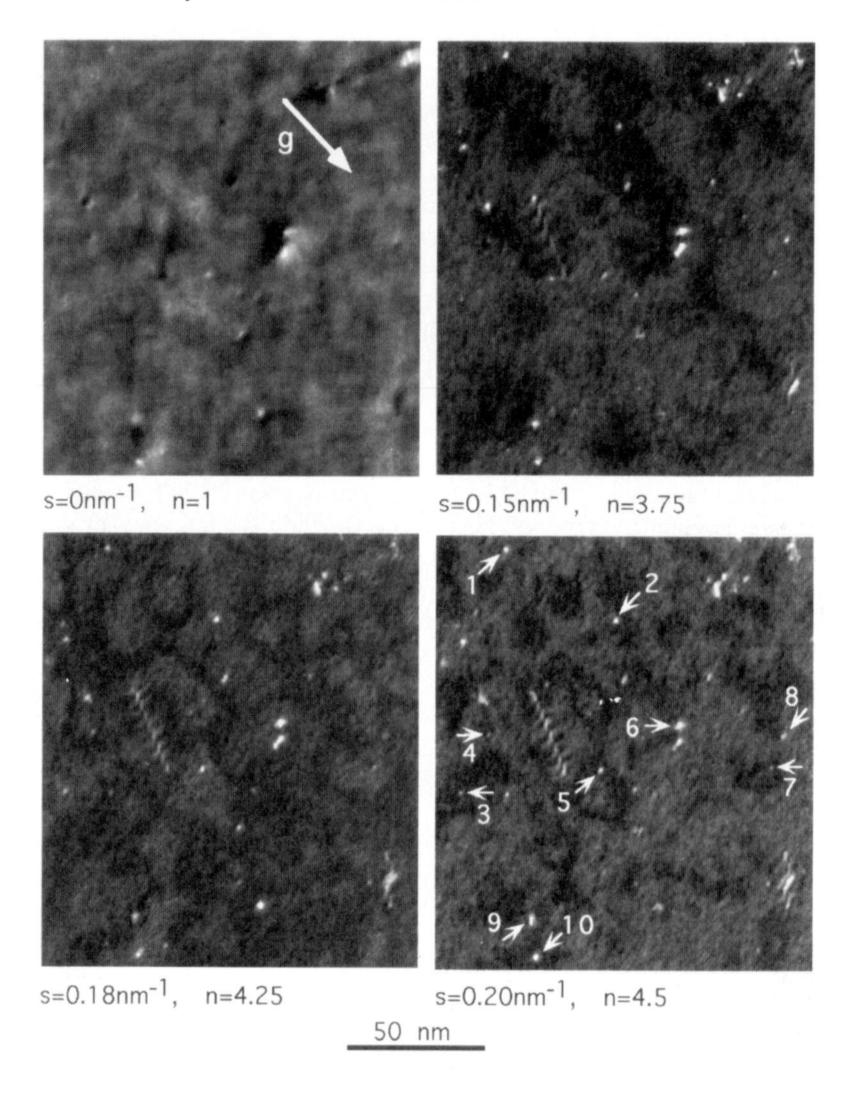

Figure 5.3. A series of weak-beam micrographs of small clusters in ion-irradiated copper, taken in $\boldsymbol{g} = 002$, $s_g > 0$, with $|s_g|$ ranging in seven approximately equal steps from 0.15–0.29 nm^{-1}. Values of n in the $(\boldsymbol{g}, n\boldsymbol{g})$ notation are also shown. The micrograph at the top left is a conventional dark-field two-beam image ($s_g = 0, n = 1$). The ten defects labelled 1–10 were analysed in detail by Jenkins *et al* (1999a), see text.

of changes in the sizes of small defects produced by a low-temperature *in situ* irradiation and annealing experiment, see section 5.2.1, example 1.

A similar approach can be used to measure the sizes of small SFT. Weak-

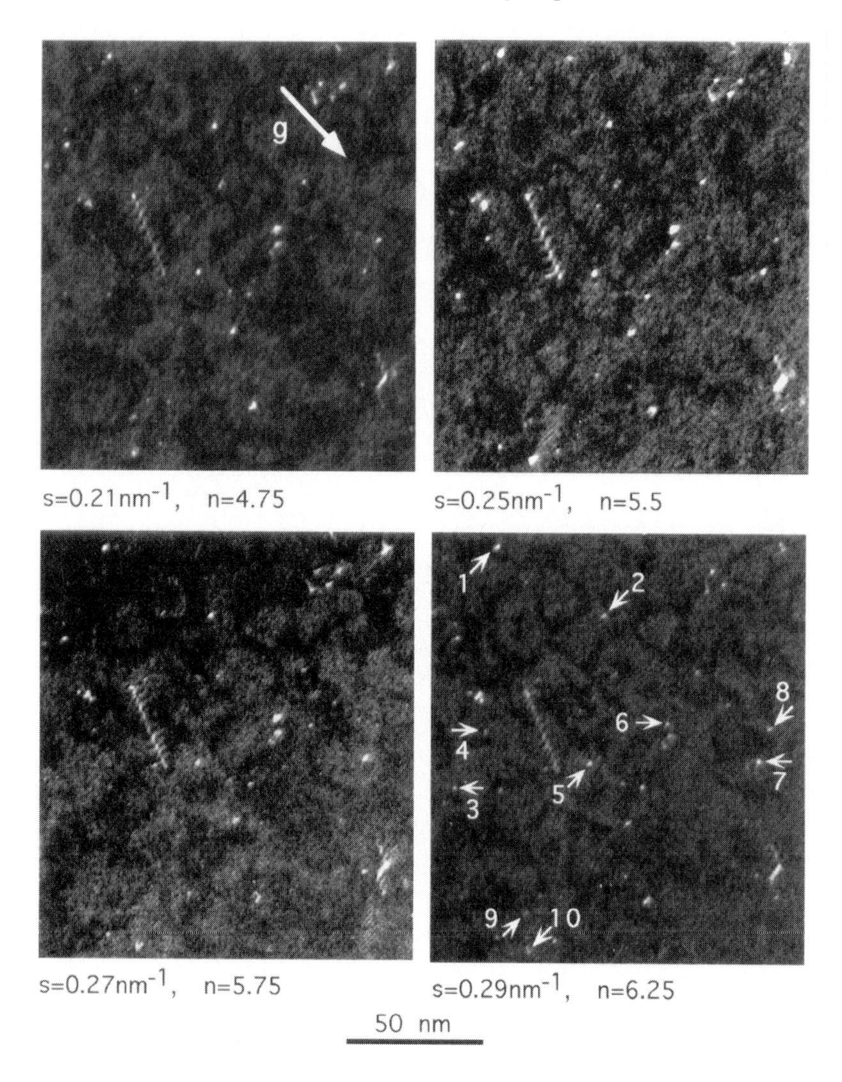

Figure 5.3. (Continued)

beam images of SFT taken using $\boldsymbol{g} = \pm 2\bar{2}0$ are shown in figure 5.5. As we have seen previously, the images of SFT under weak-beam conditions are dominated by stacking-fault fringes arising from the triangular inclined faces of fault, with the stair-rod dislocations making only a minor contribution to the total contrast. Small SFT, of size $\leq \xi_g^{\mathrm{eff}} \approx 5$ nm, generally show just one fringe from each inclined face. Somewhat larger SFT show two or more fringes. The images are generally smaller than the true SFT size, and are dependent on factors such as $|s_g|$, the foil thickness and defect depth. For different values of $|s_g|$, the fringes are displaced

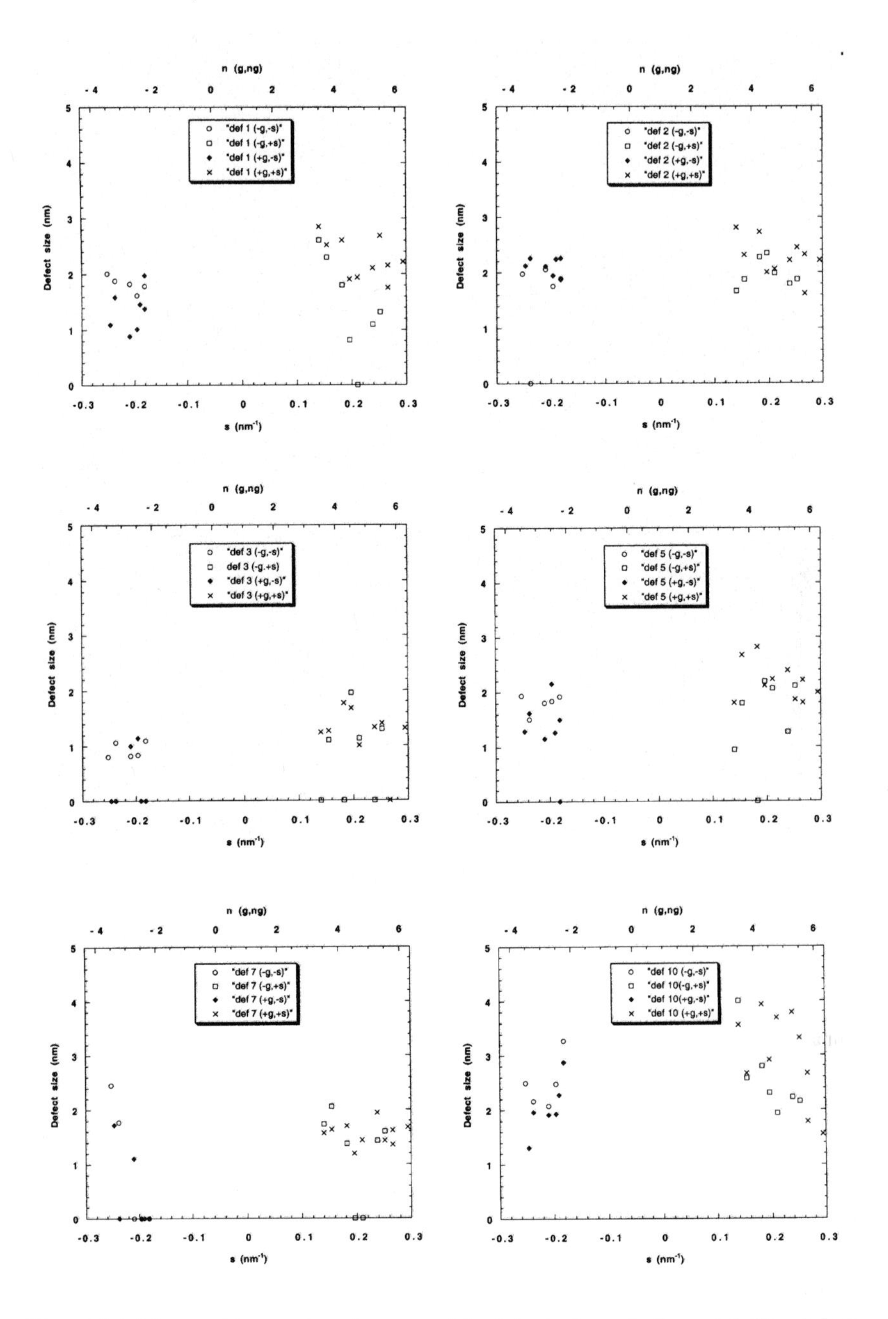

Figure 5.4. Measured image sizes for some of the defects shown in figure 5.3. The defects were imaged in $\pm \boldsymbol{g}$ for both positive and negative values of s_g. Size measurements are shown as a function of s_g. From Jenkins *et al* (1999a).

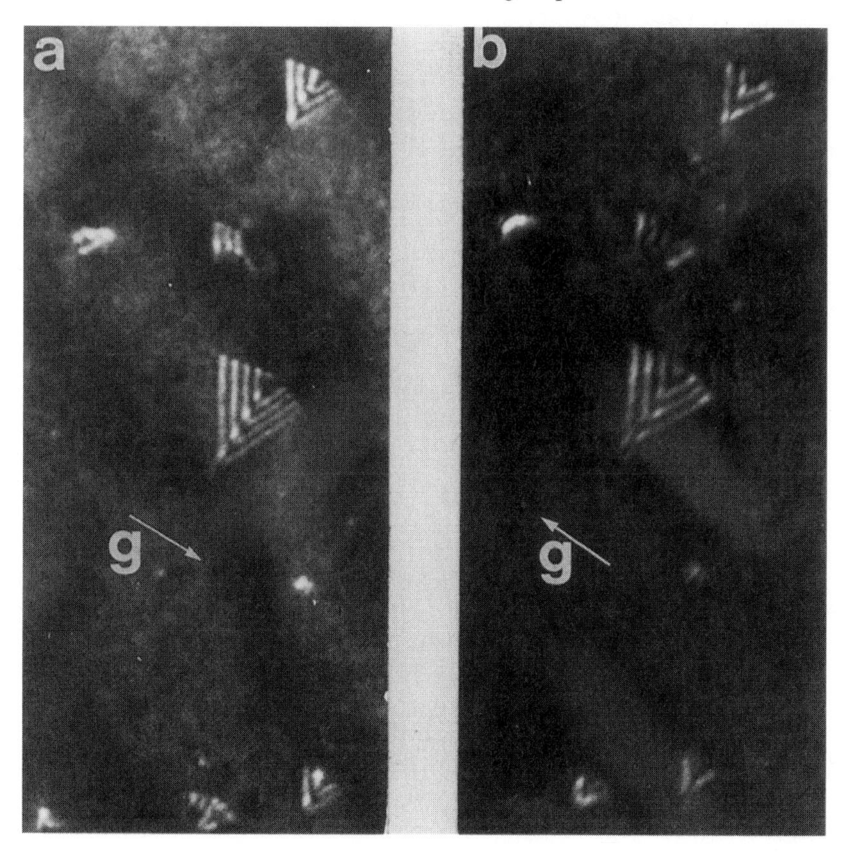

Figure 5.5. Weak-beam images of SFT in quenched gold taken in (*a*) $\boldsymbol{g} = 2\bar{2}0$ and (*b*) $\boldsymbol{g} = \bar{2}20$. The value of s_g was set at approximately 2×10^{-1} nm^{-1} in both micrographs, although the fringe spacing suggests that s_g is in fact a little larger in (*a*). Note that the number of stacking-fault fringes on the faces of the tetrahedra is sometimes larger in (*a*) than in (*b*). Because of this the apparent sizes of the smaller SFT especially are greater in (*a*) than in (*b*). Scale marker 50 nm. From Jenkins (1994).

in depth, moving up or down the face of the fault. Single weak-beam images may therefore give misleading impressions of the sizes of small SFT because the positions of the fringes on the fault faces can vary. This is clearly apparent in figure 5.5.

Satoh *et al* (1994) suggested that if micrographs of SFT were recorded for several different values of $|s_g|$, the longest fringe might be expected to correlate reasonably well to the maximum extent of the fault. The *maximum* image size should therefore be taken to correspond to the true SFT size. Experimental studies of small (~2 nm) SFT in quenched silver imaged under a wide range of $|s_g|$, from two-beam ($|s_g| = 0$) to weak-beam with $|s_g| = 0.3$ nm^{-1}, were in good accord with this hypothesis. The image sizes of *individual* SFT varied with deviation from the Bragg condition, but in each case showed a well-defined maximum. Overall, the visibility was best for $|s_g| \geq 0.2$ nm^{-1}, but the *average* image sizes of about 300 SFT was only weakly dependent on $|s_g|$.

5.2.1 Examples of image sizing by weak-beam microscopy

Example 1: The search for interstitial loops produced in displacement cascades

The possible formation of interstitial dislocation loops in displacement cascades is a subject of considerable current interest to the radiation-damage community. Molecular dynamics simulations suggest that appreciable interstitial clustering occurs in cascades (e.g. Foreman *et al* 1992, Bacon *et al* 1995), and some simulations have even predicted the production of loops large enough to be resolvable in the transmission electron microscope (De la Rubia and Guinan 1991). However, valid and direct experimental evidence for the production of interstitial loops in cascades has not yet been produced. We used the weak-beam method previously discussed to look for evidence for interstitial dislocation loops produced in collision cascades at 20 K in copper (Kirk *et al* 2000). Our experiment depended on careful measurements of the changes in the image sizes of defects in isochronal anneals. It was not successful in finding unequivocal evidence for these elusive defects (some tentative evidence was found) but clearly demonstrated that image-size measurements to an accuracy of ±0.5 nm were possible.

In situ irradiations of thin copper foils, of thickness about 60 nm, were performed with 600 keV Cu^+ ions, with the sample temperature held at 20 K in a double-tilt low temperature holder in the Argonne National Laboratory HVEM-Tandem Facility[1]. Such irradiations produce high-energy collision cascades throughout the foil, leading to the heterogeneous production of dislocation loops at cascade sites. At 20 K nucleation and growth of loops by long-range defect migration can be excluded. The improved weak-beam TEM technique described earlier was then used to measure cluster sizes recorded at 20 K before and after an isochronal anneal to 120 K. It was hoped to deduce the interstitial or vacancy nature of a cluster by its change in size following this anneal through the free interstitial migration stage I. Real size changes of the loops should be manifest in a systematic change in image sizes, measured in several micrographs taken at different values of s_g. Vacancy loops would shrink on assimilating mobile interstitials, and interstitial loops would grow.

Figure 5.6 illustrates the changes in defects in the same area before and

[1] The Argonne HVEM Tandem Facility is described in more detail by Allen and Ryan (1998). Further details are also given in chapter 9. Other *in situ* experiments carried out in this facility are described in chapters 7–9.

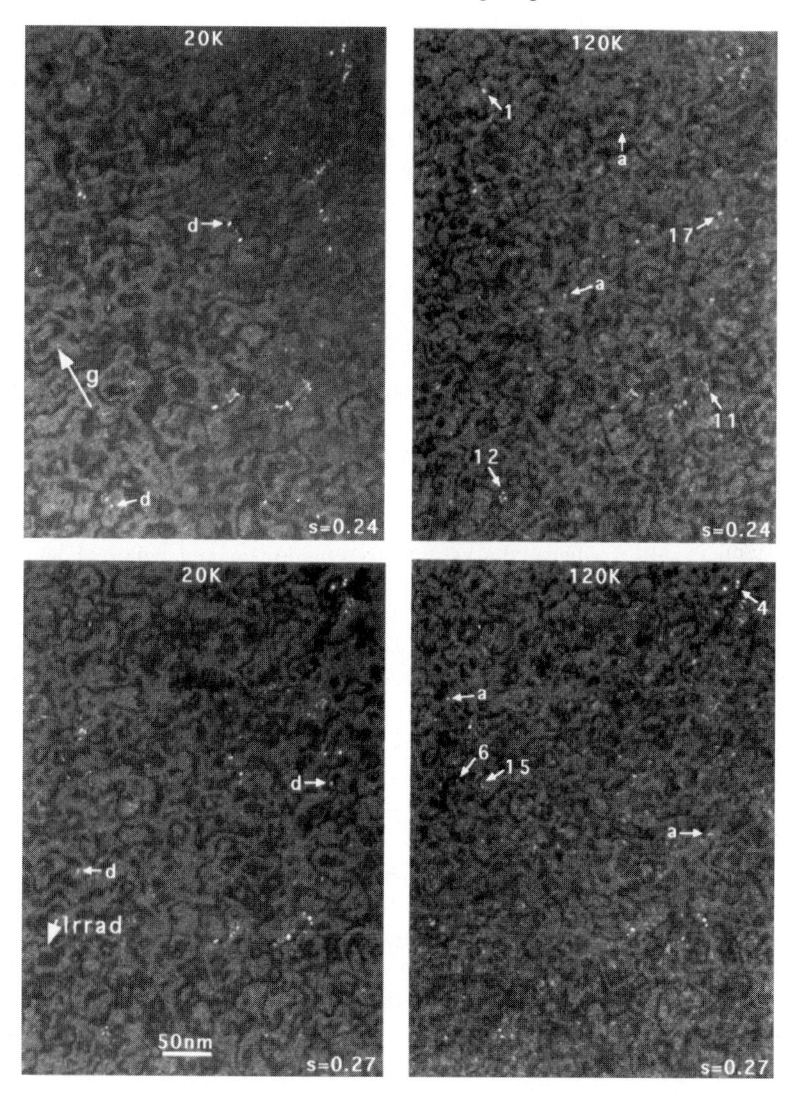

Figure 5.6. Examples of TEM micrographs of the same area of a copper foil taken after irradiation by 600 keV Cu^+ ions to a dose of 5×10^{14} m^{-2} at a sample temperature of 20 K. The micrographs on the left were taken at 20 K immediately after irradiation. The micrographs on the right were taken at 20 K after the irradiation and a 15 min anneal to 120 K. All micrographs were taken in a dark field under weakly diffracting conditions with $\boldsymbol{g} = 002$, beam direction near the [110] pole, and examples of two of seven weak-beam conditions: $s_g = 0.24$ nm^{-1} $(\boldsymbol{g}, 5.25\boldsymbol{g}$; top micrographs) or $s_g = 0.27$ nm^{-1} $(\boldsymbol{g}, 5.75\boldsymbol{g}$; bottom micrographs) are illustrated. The irradiation direction is shown with an arrow whose length is equal to the projection of the ion path through this foil thickness of 55 nm. Labelled defects are discussed in the text. From Kirk *et al* (2000).

after a 15 min anneal at 120 K. As previously indicated, the experiment required micrographs to be taken at several (in this case seven) values of s_g. Two of these seven values of s_g (in units of nm^{-1}) are shown as examples. Several clusters of defects can be seen in the left-hand images (taken before the anneal) to lie approximately along the irradiation direction and over distances equal to the projected ion path.

The pair of micrographs taken before the anneal show the small contrast differences expected when s_g is varied—individual defects show variations in contrast intensity and apparent size, and a few are visible in one micrograph but not the other—but generally they look similar. The same is true of the pair of micrographs taken after the anneal. However, comparison of micrographs taken at the same value of s_g before and after the anneal show more profound changes. This was confirmed by viewing the whole series of seven micrographs at different s_g values before and after the anneal.

Some defects of interest are arrowed and labelled. Examples of defects which disappear upon anneal are labelled 'd'. Those which appear upon anneal are labelled 'a'. Examples of defects which shrink in size are labelled '4' and '6', those which grow are labelled '1' and '12'. Defects which stay the same size are labelled '11', '15' and '17'. The directions of the arrows of numbered defects corresponds to the directions of the size measurements, which were kept constant for each defect in different micrographs.

The success of this experiment depended on there being sufficient free interstitials present after the irradiation at 20 K for measurable loop-size changes to occur when these interstitials became mobile during the anneal. Unfortunately, most clusters which persisted through the anneal behaved like '11', '15' and '17' and showed no size changes within the resolution of the weak-beam sizing technique. These clusters therefore changed in size by <0.5 nm. Of those defects whose measured size changes exceeded statistical error, the majority (such as '4' and '6') showed size decreases implying a vacancy nature. This was not unexpected: the 'collapse' of the vacancy-rich cascade core to a loop is known to occur at low temperatures (Black *et al* 1987). In addition, however, a small minority of clusters (such as '1' and '12') showed consistent image size increases and were judged, therefore, to be more likely to be interstitial in nature. The poor statistics make this result tentative.

However, perhaps more interesting aspects of this experiment were two further observations, which could only have been made by careful *in situ* weak-beam imaging. The first, most unexpected and still largely unexplained, observation was that, on annealing to 120 K, a fraction of about 25% of the clusters disappeared (such as those marked 'd' in figure 5.6), whilst a similar number of clusters ('a') appeared in different locations. Since images in several imaging conditions were available, we could exclude that this was an imaging artefact. A judgement on the nature of these clusters could not be made but vacancy collapse, interstitial clustering and complete recombination all appear possible. A video experiment suggested that loop movement is less likely.

Second, on warming specimens to room temperature a high density of small SFT appeared close to the electron-exit surface of the foil in regions which had been exposed to the electron beam at low temperatures (figure 2.12). These are most likely due to the clustering of vacancies produced by surface sputtering. Sputtering results in surface vacancies, which are immobile at 20 K. Upon warming through recovery stage III (250 K in Cu) a fraction of these vacancies migrate away from the surface with a sufficient instantaneous concentration to form clusters near the back surface. The same sputtering and vacancy migration will occur under electron irradiation at room temperature, but now the instantaneous concentration of mobile vacancies will be lower and cluster defects generally not form. However, subtle changes to existing microstructure could occur due to this process of vacancy injection by sputtering. It is interesting to note that sputtering occurred even at an electron energy of 100 keV, which is far below the threshold (about 400 keV) for displacements in the bulk.

Example 2: Matrix damage in neutron-irradiated iron

The formation of 'matrix damage' has been identified as a primary mechanism of the hardening and embrittlement of nuclear reactor pressure vessels and other near-core ferritic components, and is becoming an increasingly important factor in plant-life extension analysis. Matrix damage arises from the aggregation of irradiation-induced defects to form clusters. At present, evidence for matrix hardening is mostly indirect, coming from its effects on mechanical behaviour. It is now fairly well established that there is a hardening $\Delta\sigma_y$, which is proportional to the square root of the irradiation dose. This observation is consistent with a hardening mechanism involving the cutting by glide dislocations of irradiation-produced point-defect clusters. Several different cluster types may be involved, depending on the steel and the irradiation conditions. Suggestions for possible cluster types have included interstitial and vacancy loops; loose aggregates such as vacancy sponges; decorated microvoids; and solute point-defect clusters. However, the exact nature of such clusters has so far eluded direct characterization.

Weak-beam microscopy, possibly combined with other modern imaging and diffraction techniques such as energy-filtered imaging (see chapter 10), offers possibilities for directly revealing and characterizing matrix damage in the TEM. This assertion is supported by the following results of a study of matrix damage in relatively pure iron, neutron-irradiated to a dose of 0.06 displacements per atom (dpa) at 280 °C (Nicol *et al* to appear in the *Proceedings of the MRS Fall Conference (Boston)* December 2000). After irradiation, the material showed a significant hardness change, indicating the presence of matrix damage.

A weak-beam micrograph of the irradiated material is shown in figure 5.7. Matrix damage is visible, and consists in this case of small (2–6 nm) dislocation loops, which are visible as white dots. As emphasized in this chapter, the contrast under weak-beam conditions of an individual defect may depend on the exact

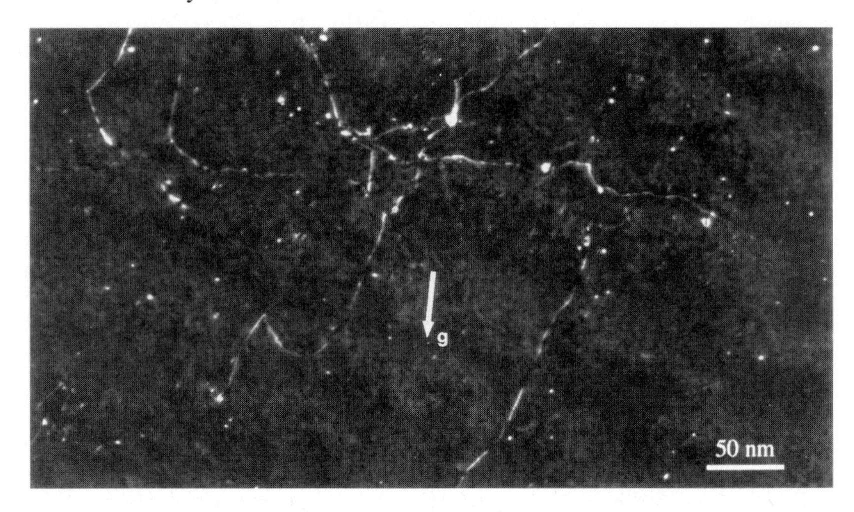

Figure 5.7. Matrix damage in relatively pure iron, consisting of small (2–4 nm) dislocation loops, which are visible in this micrograph as white dots. By recording images at several different values of s_g (as discussed in the text) it is possible to reliably identify and size dislocation loops as small as 1–2 nm.

imaging conditions (e.g. the depth in the foil, and foil thickness, as well as the deviation parameter s_g) but by recording images at several different values of s_g it was possible to reliably identify and size dislocation loops as small as 1–2 nm. The loops in this case have a mean image size d of 4.2 ± 0.3 nm.

Weak-beam microscopy was also capable of determining the loop Burgers vectors $\boldsymbol{b}$ (as described in chapter 3). Figure 5.8 shows the same area of foil imaged (*a*) with diffraction vector $\boldsymbol{g} = 1\bar{1}0$, and (*b*) $\boldsymbol{g} = 002$. In figure 5.8(*a*) two of the three loop variants with $\boldsymbol{b} = \langle 100 \rangle$, namely [100] and [010], should show strong contrast. In figure 5.8(*b*) these two loop variants would have $\boldsymbol{g} \cdot \boldsymbol{b} = 0$, and so are out of contrast. Looking at figure 5.8, it is immediately obvious that many more defects are visible in (*a*) than in (*b*). A more complete contrast analysis does indeed confirm that about 80% of the loops have Burgers vectors $\boldsymbol{b} = \langle 100 \rangle$, whilst the remainder have $\boldsymbol{b} = \frac{1}{2}\langle 111 \rangle$. Estimates of the loop number density N_v gave a value of about 8.5×10^{21} m^{-3}.

The vacancy or interstitial nature of the loops was not determined directly in this experiment. However, several factors point to an interstitial nature for at least some of the loops. First, an *in situ* annealing experiment showed that loops were stable to temperatures of at least 430 °C. Vacancy loops would be expected to shrink quickly by vacancy emission at this temperature. Second, in experiments in a somewhat purer alloy a lower loop density was found, suggesting a role of impurities in loop nucleation, which would be more likely for interstitial loops. Third, it is known that vacancy loop production does not usually occur in cascades

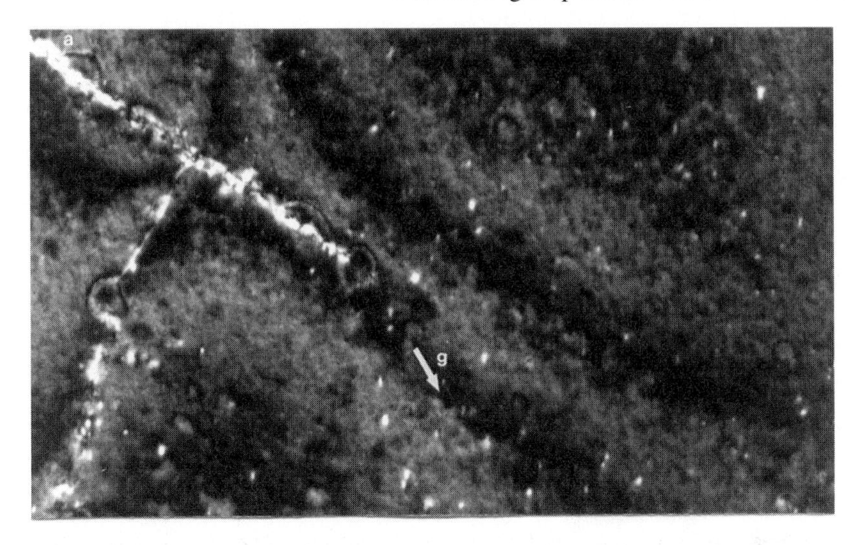

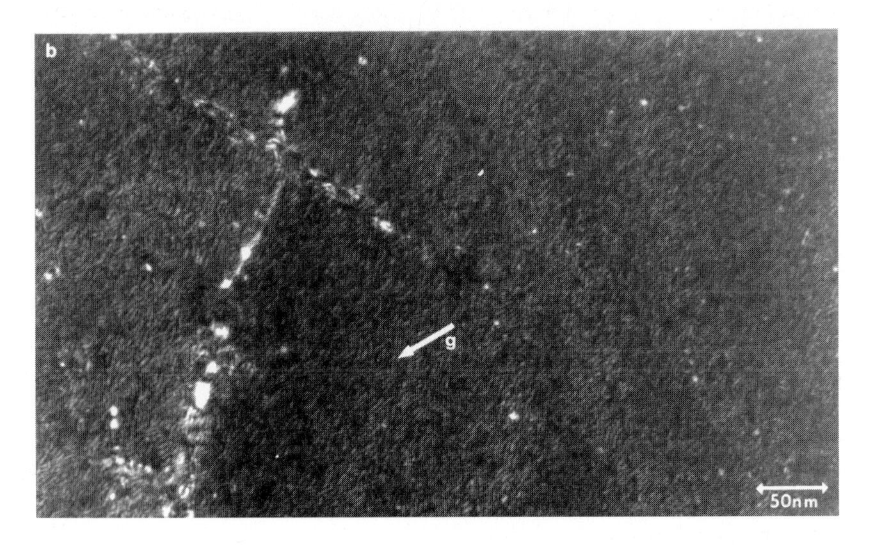

Figure 5.8. The same area of foil in relatively pure iron imaged (*a*) with diffraction vector $\boldsymbol{g} = 1\bar{1}0$ and (*b*) $\boldsymbol{g} = 002$. In (*a*) two of the three loop variants with $\boldsymbol{b} = \langle 100 \rangle$, namely [100] and [010], show strong contrast. In (*b*) these two loop variants have $\boldsymbol{g} \cdot \boldsymbol{b} = 0$, and so are out of contrast, which accounts for the smaller number of loops visible in this micrograph.

in iron initiated by iron primary knock-ons. Fourth, in experiments carried out by Hoelzer and Ebrahimi (1995) in a similar material, irradiated under similar conditions to the present study, some of the loops were sufficiently large that their nature could be shown to be interstitial by the inside–outside contrast technique (see chapter 4).

From the measured loop sizes and number densities, it is possible to estimate the hardening produced by this population of loops. The calculated loop hardening values were found to be somewhat less than the observed values. This suggests that there is some additional hardening contribution from defects not seen in the weak-beam micrographs.

Chapter 6

Characterization of voids and bubbles

This chapter is concerned with methods of analysing cavities—voids and bubbles. Voids are three-dimensional aggregates of vacancies. In most cases voids are stabilized by the presence of gases such as hydrogen or helium, in which case they are more properly called bubbles. Voids and bubbles occur in a wide range of materials irradiated with fast neutrons at high temperatures, and can have severe deleterious effects on reactor components. For example, voids can cause appreciable swelling and distortion of stainless steel reactor components, a phenomenon known as 'void swelling'. Equilibrium bubbles and large empty or partially-filled voids generally have no significant long-range elastic strain field. For convenience we will refer to all subsurface cavities with negligible strain fields as voids, whereas cavities with appreciable strain fields will be termed bubbles, although this distinction is somewhat arbitrary.

6.1 'In-focus' imaging of larger voids

Voids of diameter d larger than about 5 nm may be imaged by structure factor contrast under dynamical or bright-field kinematical imaging conditions. The contrast mechanism is similar to that for disordered zones in ordered alloys or amorphous zones in crystalline matrices (section 2.2 and chapter 7). Effectively, a column passing through a void is shorter than an adjacent column passing through perfect material. It follows that voids imaged by this mechanism are best visible when viewed in-focus under dynamical conditions in the flanks of low-order thickness fringes. They will appear dark at the edge of a bright fringe, because here the intensity increases rapidly with increasing foil thickness, and bright at the edge of a dark fringe because here the intensity decreases rapidly with increasing foil thickness (see figure 2.9). Large faceted voids may show thickness fringes at some foil orientations because the foil is effectively wedge-shaped. With increasing foil thickness, as thickness fringes die out due to anomalous absorption, the contrast of voids becomes weaker and they always appear lighter than background. In the literature, larger voids usually show images of this type.

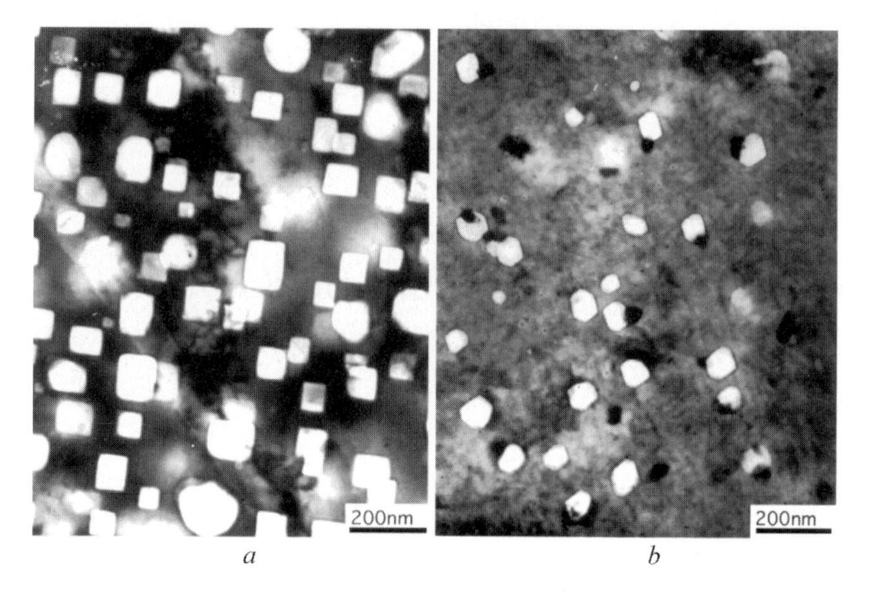

Figure 6.1. In-focus image of faceted voids in high-dose neutron-irradiated stainless steels containing different amounts of Ti: (*a*) 16Cr–15Ni–3Mo–0.3Ti, and (*b*) 16Cr–15Ni–3Mo–1Ti. Specimens were irradiated to a dose of 60 dpa at 480 °C. Void swelling is considerably less in the steel containing the higher Ti content, evidenced by the lower void density and smaller void sizes in (*b*). Note that voids are often associated with precipitates, identified as Ni_3Ti. After Sagaradze *et al* (1999).

Examples of large voids imaged in thicker regions of foil are shown in figure 6.1. Voids will not be seen in in-focus images if the foil thickness t is greater than about $20d$.

In-focus imaging of voids has been used extensively to investigate void microstructures in irradiated materials. Images are often obtained in a bright field under unspecified kinematical diffraction conditions, with $w = s_g \xi_g \neq 0$. However, the contrast mechanism remains an effective change in foil thickness. Under such kinematical conditions, the contrast of voids may be lighter or darker than background, depending on void depth and the sign and magnitude of w. The contrast of voids smaller than about $\frac{1}{3}\xi_g$ is so weak that they are unlikely to be visible.

6.2 'Out-of-focus' imaging of smaller voids

Smaller voids are often best imaged in an out-of-focus imaging condition, and this may be the only way to image voids smaller than about 5 nm. The contrast arises from a weak absorption component and a phase-contrast component due to defocus. The phase-contrast component arises from the change in mean inner

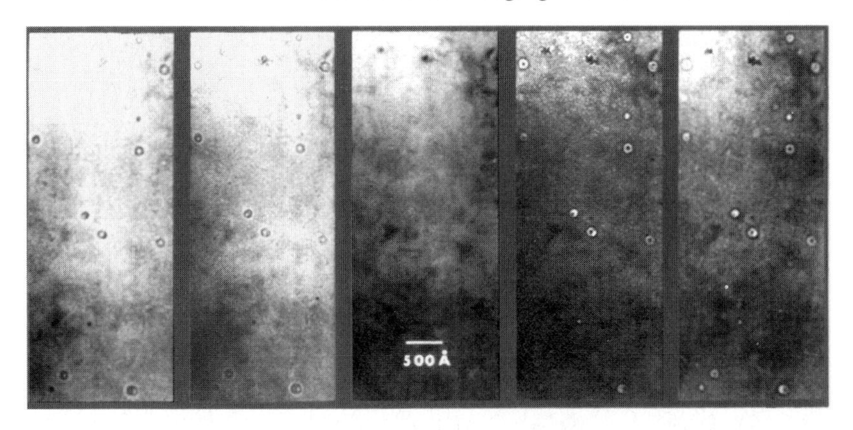

Figure 6.2. A through-focal series of small voids in stainless steel irradiated with 3 MeV Ni^+ ions, underfocus on the right-hand side, and overfocus on the left. The total change in focus Δf is about 300 nm. Voids appear as predominantly white dots surrounded by a dark Fresnel fringe in the underfocus images, and as predominantly dark dots surrounded by a bright fringe in the overfocus images. The micrograph at the centre was taken at the in-focus condition, when the voids are not visible (from Rühle 1972).

potential between the void and the matrix, which causes a phase shift between electrons which traverse the void and those which pass through adjacent perfect crystal. This gives rise to Fresnel fringes near the edge of the void. The effect depends sensitively on the degree of under- or overfocus of the objective lens. In order to confirm the presence of small voids, a bright-field through-focal series should be recorded with the foil tilted well away from the Bragg condition for all reflections. Under such kinematical conditions, voids appear as white dots surrounded by a dark Fresnel fringe in underfocus images, and as dark dots surrounded by a bright fringe in overfocus images. Better contrast is usually obtained in underfocus images, which correspond to the objective lens being weaker than the in-focus condition. An example of such an experiment is shown in figure 6.2. Note that in the in-focus micrograph the voids are not visible. Voids as small as 1 nm can be detected using defocus values of about 1 μm.

Through-focal images of voids are very characteristic, but it may in some cases be necessary to carry out stereo measurements to distinguish small voids from surface craters which show essentially the same contrast. Examples of craters in molybdenum produced by energetic molecular ion irradiation are shown in figure 6.3. These craters are believed to be produced by the intersection of displacement cascades with the foil surface (English and Jenkins 1987; see also section 9.3). Note that the contrast is indistinguishable from that shown by subsurface cavities.

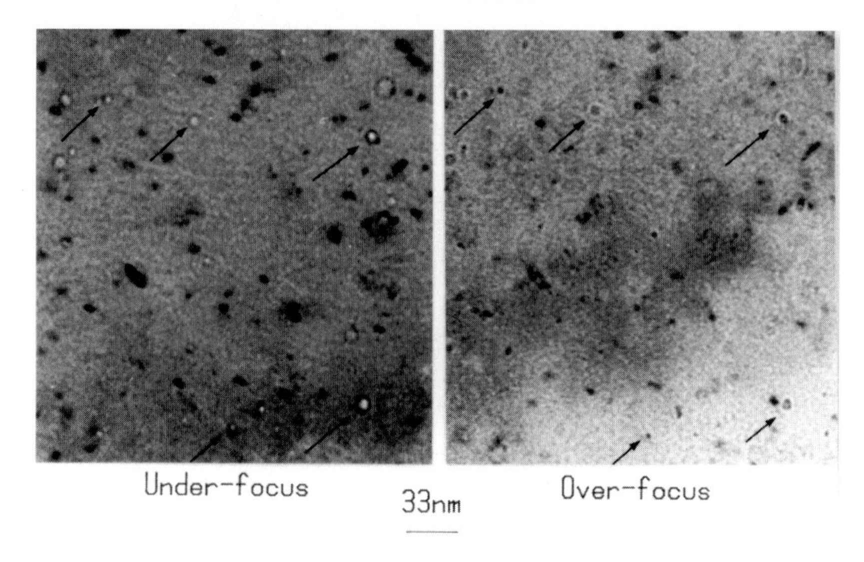

Figure 6.3. Surface craters in molybdenum, produced by irradiation with 180 keV Sb_3^+ molecular ions at room temperature. The same area is shown, imaged in bright-field conditions in both underfocus and overfocus conditions, with $\boldsymbol{g} = 04\bar{4}$ and with s_g positive. Some craters are arrowed. They show phase contrast behaviour typical of cavities. Micrograph courtesy of Dr C A English.

6.3 Sizing of voids

For in-focus images of large voids, calculations by Rühle (1967) indicate that the outer diameter of the contrast figure corresponds well to the actual size of the void.

Simulations of out-of-focus images of voids have also been made by the Stuttgart group. Rühle and Wilkens (1975) developed the necessary theory and applied it to calculate the contrast of spherical voids imaged under two-beam conditions, close to a Bragg condition. Their main results for small voids in thick foils are shown in figures 6.4 and 6.5. In figure 6.4 the intensity relative to background, I/I_0, is shown as a function of the reduced radial distance $\rho = r/r_0$, where r is the distance from the centre of the void. The calculations are made for three different values of underfocus, defined by the parameter $\beta = \Delta f/\pi k_0 r_0^2$ (where k_0 is the electron wavevector λ^{-1}, Δf is the defocus distance and r_0 is the radius of the void). In all cases, the intensity is highest at $r = 0$, and then drops with a steep gradient below background at some distance r_c which does not necessarily coincide with r_0. The resulting image is a white dot surrounded by a series of dark and bright Fresnel fringes, from which the first ring should be resolved. For values of defocus of the order of -1 μm, the diameter of the inner bright area (corresponding to the inner diameter of the first dark Fresnel fringe,

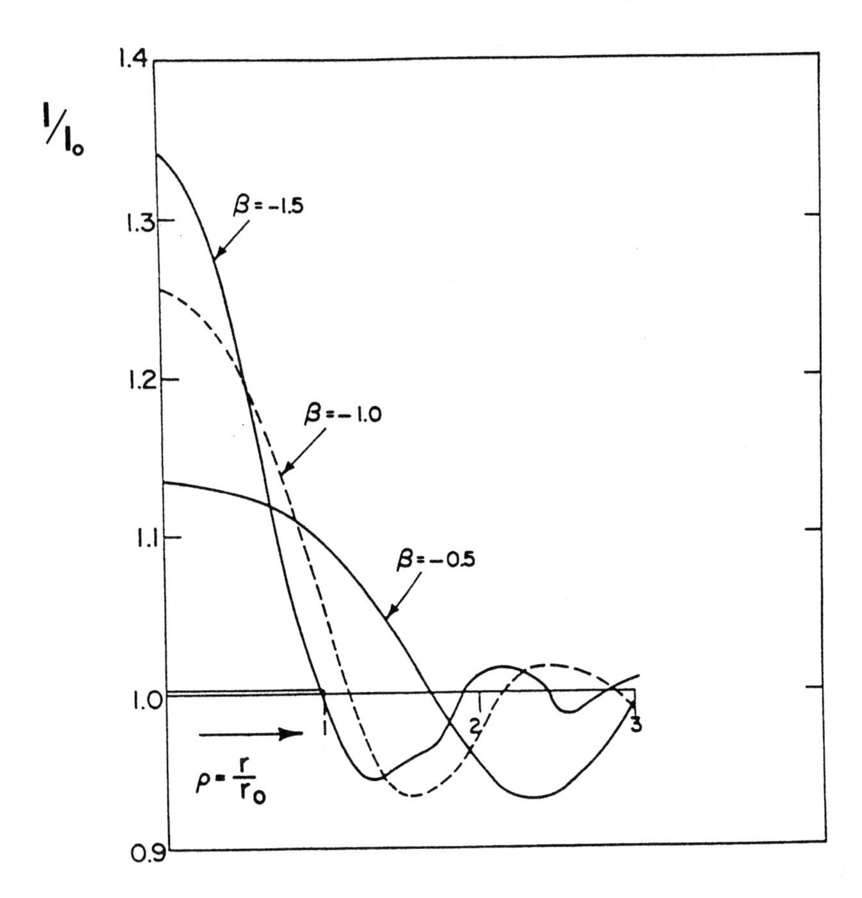

Figure 6.4. Calculations from Rühle and Wilkens (1975) of the 'out-of-focus' contrast of small spherical voids in a thick foil, in the underfocused condition. The intensity relative to background I/I_0 is shown as a function of the reduced radius $\rho = r/r_0$ where r_0 is the void radius, for different values of defocus. Here, $\beta = \Delta f/\pi k_0 r_0^2$, where k_0 is the electron wavevector λ^{-1} and Δf is the defocus distance.

d_{in}) corresponds within 10% to the actual void size for voids larger than about 2 nm. This is shown in figure 6.5 for $\Delta f = -0.8$ μm.

The inside edge of the first dark fringe therefore delineates the edge of larger, spherical voids reasonably well in this imaging condition. The outer diameter of the first dark Fresnel fringe (d_{out}) may be much bigger than the void size.

The calculations summarized in the previous paragraph and figure 6.5 have led many workers to take the inside edge of the dark fringe in an underfocused image to be the perimeter of the void (see, e.g., Hertel and Rühle 1979). This choice is not necessarily optimal in all cases. It may be criticized on two

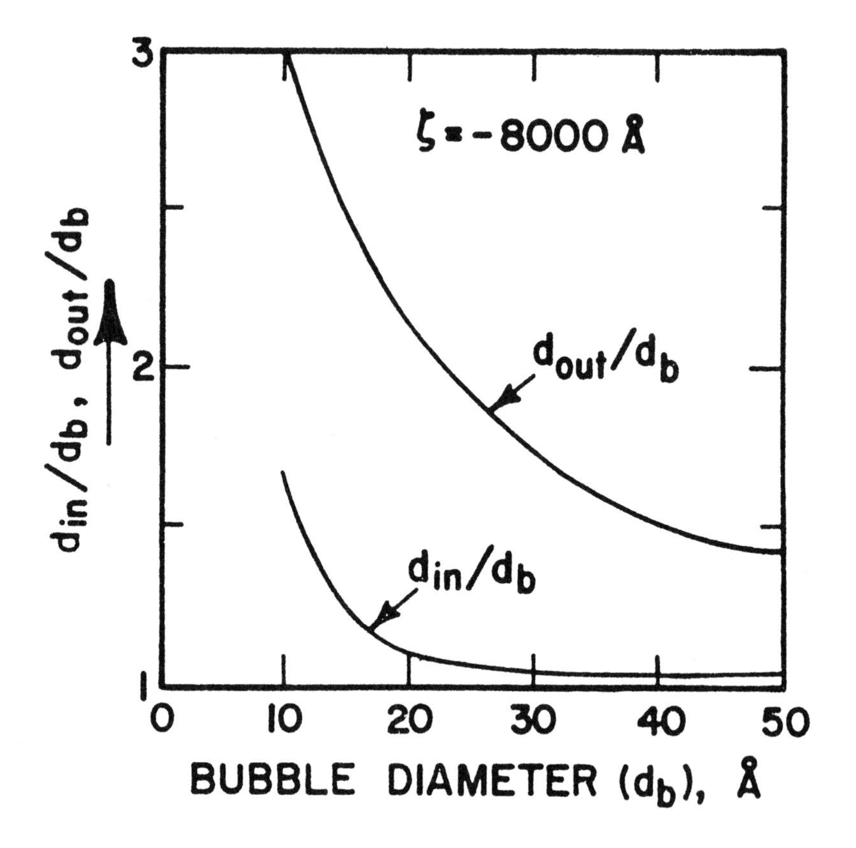

Figure 6.5. Calculations of the 'out-of-focus' contrast of small spherical voids in thick foils from Rühle and Wilkens (1975). The reduced diameters (d_{in}/d_b) and (d_{out}/d_b) of the inner and outer first dark Fresnel fringe are shown as a function of the actual void diameter (d_b) for a defocus of Δf (designated ζ by Rühle and Wilkens) $= -0.8$ μm.

grounds. First, voids are usually imaged under kinematical rather than dynamical diffraction conditions. The contrast of the voids tends to be better, and features such as dislocations are less dominant. Second, in many materials voids are not spherical but faceted. Use of the 'inside-edge of the dark fringe' criterion was thought by von Harrach and Foreman (1980) to be responsible for inconsistencies in the measured and calculated volumes of *faceted* voids in irradiated 316 stainless steel. For this reason, Foreman *et al* (1980, 1982) used the theory of Rühle and Wilkens (1975) to calculate images of spherical and faceted voids for orientations remote from strong Bragg reflections.

The main conclusions of these simulations were as follows.

(i) For spherical voids imaged, the same general contrast behaviour reported

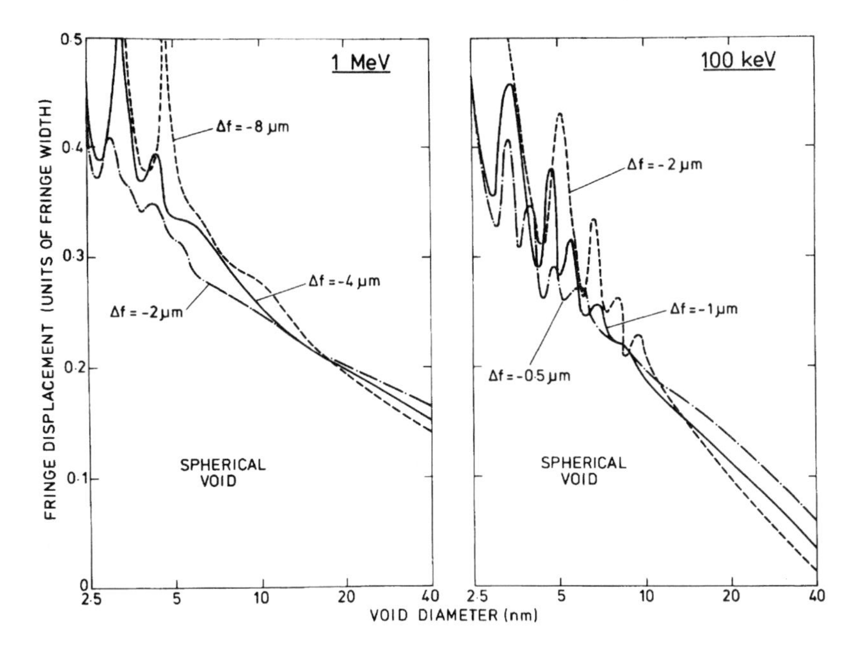

Figure 6.6. Calculations by Foreman *et al* (1982) showing the calculated outward displacement of the centre of the dark fringe from the edge of a spherical void, imaged under kinematical conditions. Note the insensitivity to defocus Δf, and the tendency of the fringe to become centred on the void edge for large voids. Electron beam energies 100 keV and 1 MeV.

by Rühle and Wilkens was found. In particular, the principal dark fringe in an underfocused image lay mostly outside the void edge for diameters $d < 5$ nm, but tended to move in as the void size increased. For void sizes greater than about 10 nm the *centre* of the dark fringe delineated the edge of the void to an accuracy better than 0.2 nm. The fringe position was also insensitive to defocus. This is shown in figure 6.6.

(ii) The same general behaviour was also found for faceted voids. The principal dark fringe again lay outside the void edge at very small void sizes. For a void diameter of $d = 2.5$ nm, the inner edge of the dark fringe defined the void perimeter fairly well. However, for larger voids the fringe moved inwards towards the void edge, particularly if the side face of the void was tilted relative to the electron beam by an appreciable angle (θ). The *centre* of the dark fringe became centred on the void edge when the linear tilt of the side face exceeded the fringe width. This is illustrated in figure 6.7, which shows calculated intensity profiles for a 20 nm faceted void with hexagonal section (with $\theta = \pi/6$, see the insert at top right). Note the dark fringe accurately centred on the void edge ($r/r_0 = 1$). Although the

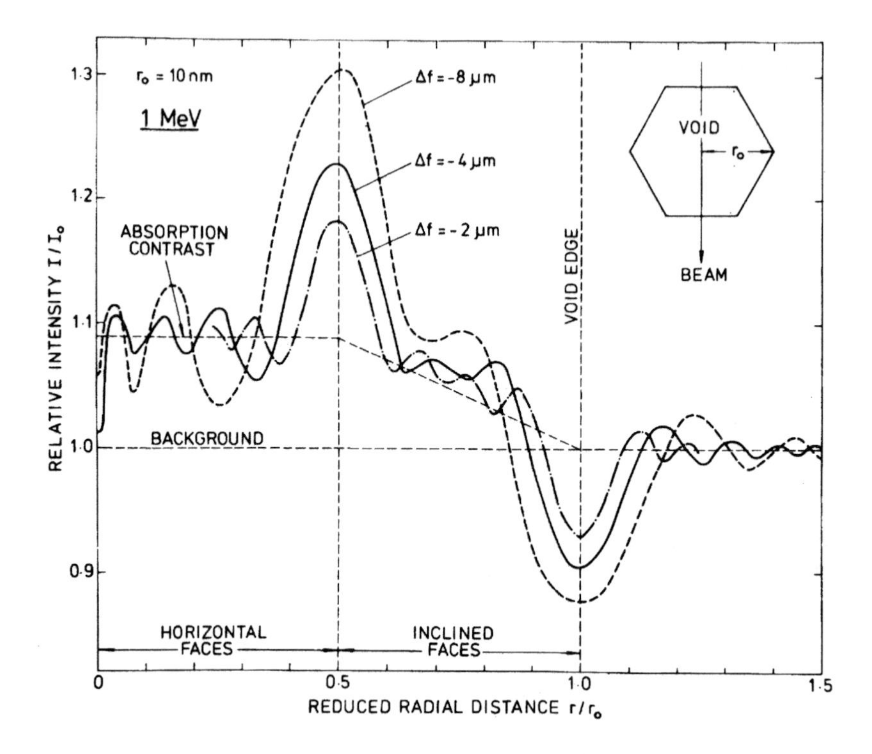

Figure 6.7. Calculations by Foreman *et al* (1982) showing calculated intensity profiles for a 20 nm diameter faceted void with hexagonal section (see insert at top right), imaged under kinematical conditions. Note the dark fringe centred on the void edge, and the bright fringe positioned at the inner edge of the inclined face. The positions of both fringes are insensitive to Δf. Electron beam energy 1 MeV.

width and intensity of this fringe are sensitive to defocus Δf, the position of this fringe is very insensitive to Δf (and other calculations showed it also to be insensitive to void size, orientation, absorption coefficient and mean crystal potential). Note also the bright fringe positioned at the inner edge of the inclined face. This behaviour differs significantly from a spherical void, where the bright fringe lies immediately inside the dark fringe.

(iii) The previous result (ii) suggests that the *centre* of the dark fringe in an underfocused image taken under kinematical conditions provides the best representation of the void perimeter for faceted voids larger than about 5 nm. Figure 6.8 shows the measured and predicted change in the apparent radius of an 80 nm diameter void with defocus distance, using either the centre or inner edge to define the void perimeter. The experiment confirmed that the apparent size, measured with the fringe-centre criterion, remained constant with defocus. However, the apparent radius measured by the inner-edge

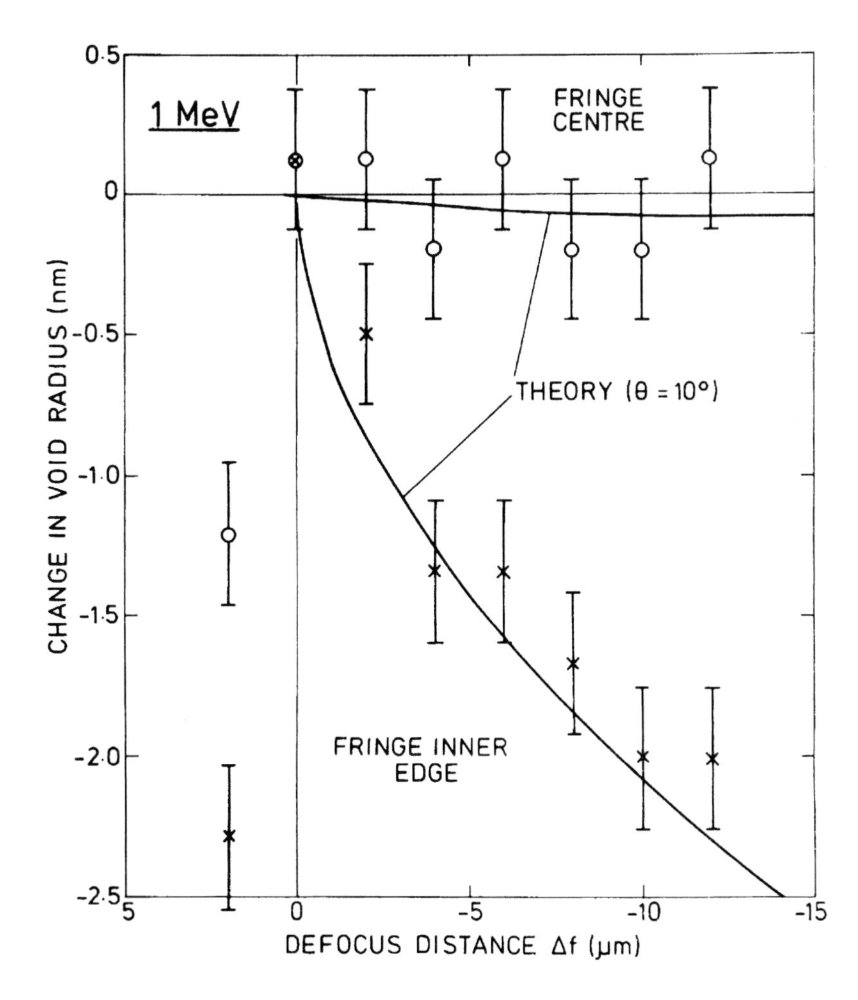

Figure 6.8. Comparison between calculated and measured changes in the radius of a faceted void with defocus, using either the dark fringe centre or inner edge. The (111) side face of the void (80 nm diameter) is tilted by $\theta = 10^\circ$. There is good agreement between theory and experiment. From Foreman *et al* (1982).

criterion decreased with defocus. In both cases there is excellent agreement with theory for $\theta = 10^\circ$, which is the inclination of the side faces of this void. It is clear that the use of the inner-edge criterion for faceted voids generally gives an underestimate of the true void size. It was estimated that this could lead to errors in the void volume of up to a factor of two for small faceted voids.

In conclusion, the best estimate of void size for voids imaged under kinematical conditions is generally obtained from the position of the centre of the principal dark fringe in an underfocused image. Note, however, that both the calculations of Rühle and Wilkens (1975) and Foreman *et al* (1982) imply that it is difficult to measure accurately the sizes of voids smaller than about 5 nm, and voids smaller than about 2 nm *cannot* be sized accurately.

6.4 Weak-beam imaging of small voids

Surprisingly, small voids are sometimes seen well under weak-beam imaging conditions. This can be particularly useful if the dislocation density in the specimen is high, such as in the example shown in figure 6.9 taken from Lee *et al* (1985). This figure shows a typical microstructure of alumina after 4 MeV Ar^+ ion irradiation to high doses at temperatures from 700–900 °C. The very high dislocation density necessitated the use of weak-beam microscopy. Note the presence of an ordered array of small voids, visible as white dots (see also section 6.6). The micrograph was taken in-focus, and there was no evidence of strain associated with the voids. The contrast mechanism therefore seems to be structure-factor contrast, the same as for in-focus imaging of larger voids. We would expect the mechanism to operate effectively if the void size is an appreciable fraction of the effective extinction distance, ξ_g^{eff}, which is about 5 nm under weak-beam conditions. This does not seem to have been explored.

6.5 Over-pressurized bubbles

Over-pressurized bubbles or voids with impurity segregation at their surfaces may have considerable strain fields. Such bubbles can be imaged using standard strain-contrast techniques in the in-focus condition as described in chapter 3. They would be expected to show similar contrast to a misfitting inclusion. The types of contrast depend on the size of the bubbles and the amount of strain they produce. This has been investigated systematically by Chik *et al* (1967). Their results are summarized in figure 6.10. Here the types of contrast shown by spherical bubbles under two-beam dynamical imaging conditions are shown as a function of the reduced void radius R_v/ξ_g and of the normalized misfit parameter P. The diagram is divided into three regions:

- Region A corresponds to small bubbles with moderate or large misfit parameters. In relatively thin foils these would show black–white contrast. For spherical voids in an isotropic material, the black–white vector $\boldsymbol{l}$ would always be parallel to $\boldsymbol{g}$, and a layer structure similar to that for small loops would be expected (see figure 3.2). The small He^+ bubbles in gold seen in figure 6.11 show contrast of this type.
- In region B, corresponding mostly to larger bubbles with large misfit parameters, black–white contrast is still observed, but the depth oscillations of the contrast are suppressed due to surface relaxation in the same manner

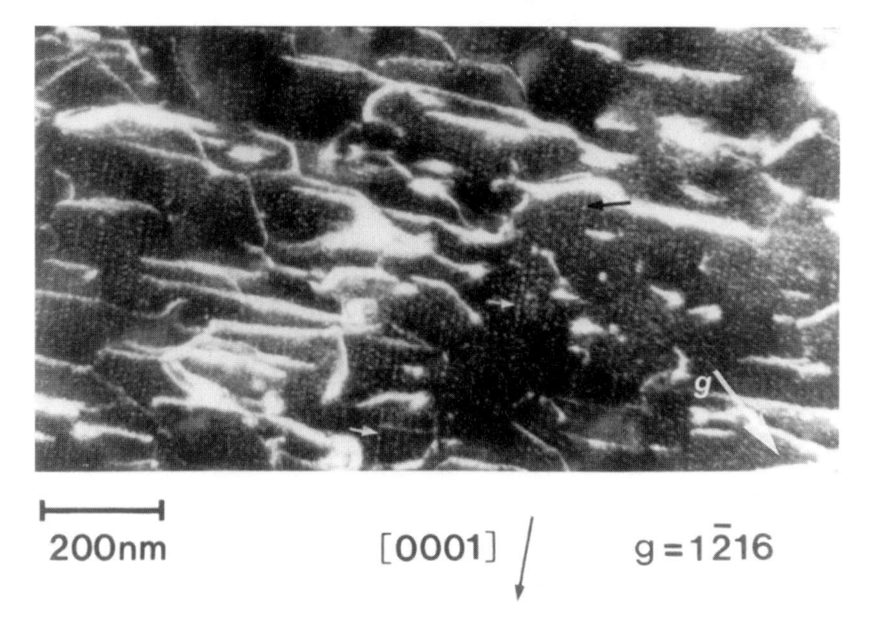

Figure 6.9. Small voids ($d < 6$ nm) visible as white dots under weak-beam imaging conditions in alumina irradiated with 4 MeV Ar^+ ions to a dose of 6 dpa at 800 °C. Some alignment along the c-axis can be seen. From Lee *et al* (1985).

as for large misfitting inclusions (Ashby and Brown 1963; see figure 3.1). $\boldsymbol{l}$ is again parallel to $\boldsymbol{g}$ with the direction of black–white streaking the same as for small bubbles located in layer L1.

- In region C, corresponding mostly to large bubbles with low or moderate misfit parameters, no black–white contrast occurs. The bubbles show pure thickness contrast as for large voids (section 6.1).

Figure 6.12 shows electron micrographs of overpressurized bubbles in electron-irradiated stainless steel observed in dynamical dark-field and bright-field conditions with $\boldsymbol{g} = \bar{2}\bar{2}0$ and in kinematical conditions with no reflection close to excitation. The dark-field image, taken in a dark extinction contour, shows black–white images with $\boldsymbol{l}$ parallel to $\boldsymbol{g}$. The bubbles labelled '2' and '3' were found by stereo microscopy to be situated within the depth layers L2 and L3. As expected for bubbles in the contrast regime A, they have opposite signs of $\boldsymbol{g} \cdot \boldsymbol{l}$. Since $\boldsymbol{g} \cdot \boldsymbol{l} < 0$ in layer L2 and $\boldsymbol{g} \cdot \boldsymbol{l} > 0$ in L3, there is a negative volume misfit. The bright-field image, taken in a bright extinction contour, shows black–black contrast very reminiscent of the contrast of misfitting inclusions (Ashby and Brown 1963; see figure 3.1). Similar double-lobe black contrast also occurs slightly away from the Bragg condition ($s_g \neq 0$). Calculations indicate that the extent of the black–white or black lobe contrast may be much larger than the

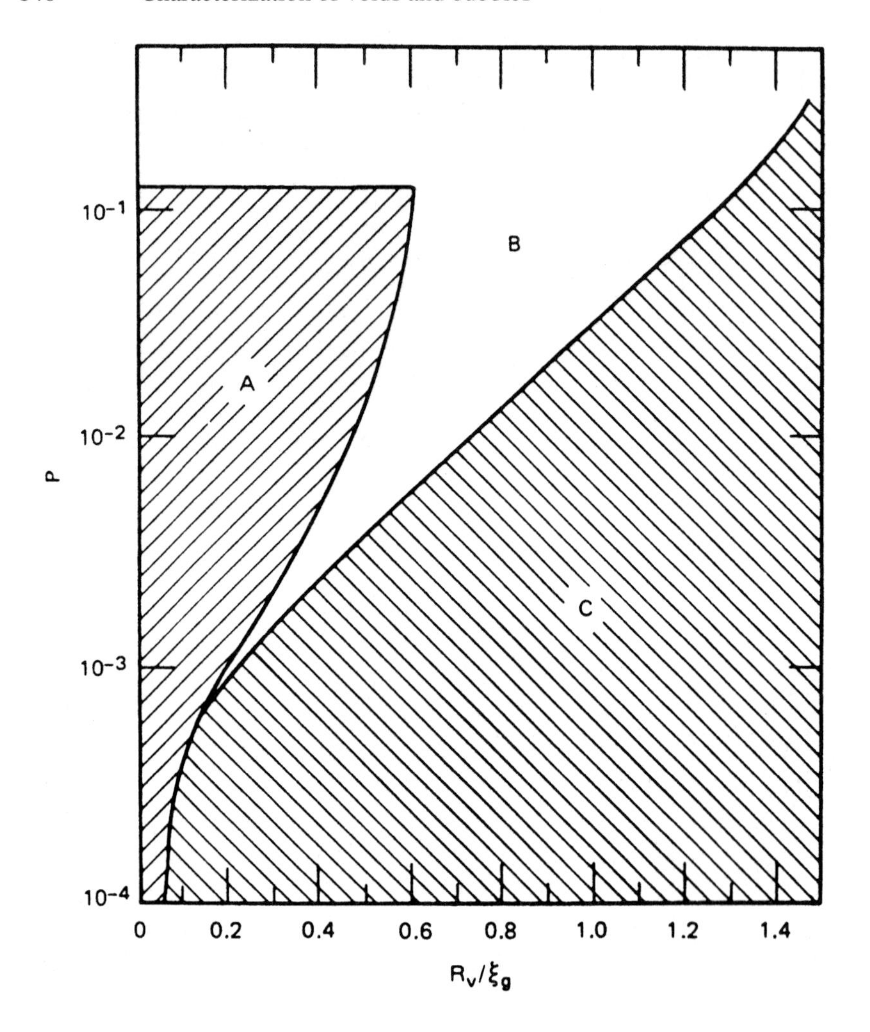

Figure 6.10. Types of contrast shown by spherical bubbles imaged under two-beam dynamical diffraction conditions in the in-focus position. The contrast types are shown as a function of the reduced void radius R_V/ξ_g and the normalized misfit parameter P. P is defined by: $P = \Delta V(|\boldsymbol{g}|/\pi\xi_g^2)$ where ΔV is the volume misfit of the defect. Region A: black–white contrast with depth oscillations. Region B: black–white contrast without depth oscillations. Region C: no black–white contrast, pure thickness contrast. From Chik *et al* (1967).

bubble itself, and this is confirmed by figure 6.12(c) where the bubbles are visible but their strain contrast is suppressed. More details can be found in Saldin and von Harrach (1980).

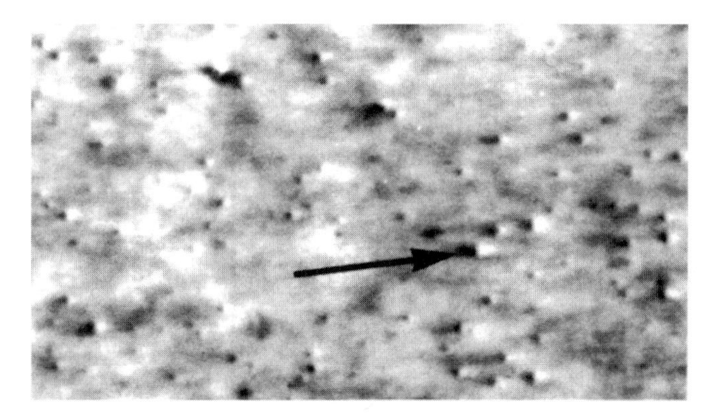

Figure 6.11. Black–white contrast from small helium bubbles in He^+ bombarded gold. The diffraction vector $\boldsymbol{g} = 220$ is indicated by the arrow. For defects such as these with symmetrical strain fields, the $\boldsymbol{l}$-vector runs approximately parallel to $\boldsymbol{g}$. The foil orientation is close to [001]. Micrograph courtesy of Professor M Rühle (1972).

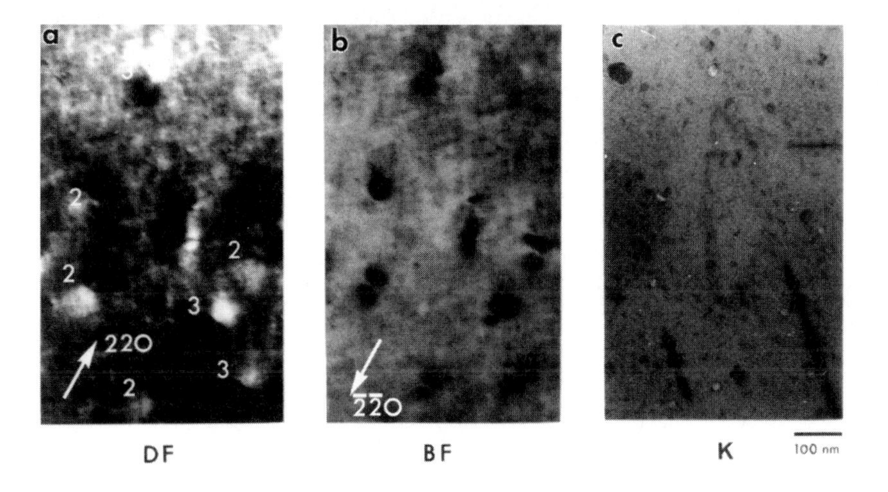

Figure 6.12. Voids with associated strain fields in electron-irradiated stainless steel, imaged (*a*) in dynamical dark-field conditions with $\boldsymbol{g} = 220$ showing black–white strain contrast with $\boldsymbol{l}$ parallel to $\boldsymbol{g}$; (*b*) in bright-field conditions, with $\boldsymbol{g} = \bar{2}\bar{2}0$, showing double-lobe contrast; and (*c*) in kinematical conditions with no reflection close to excitation. Note that the extent of the strain contrast is much larger than the size of the void causing it. From Saldin and von Harrach (1980).

6.6 Void arrays

Under some irradiation conditions, periodic ordered arrays of voids or bubbles are formed in some metals and ceramics. Such arrays may be one-, two- or three-

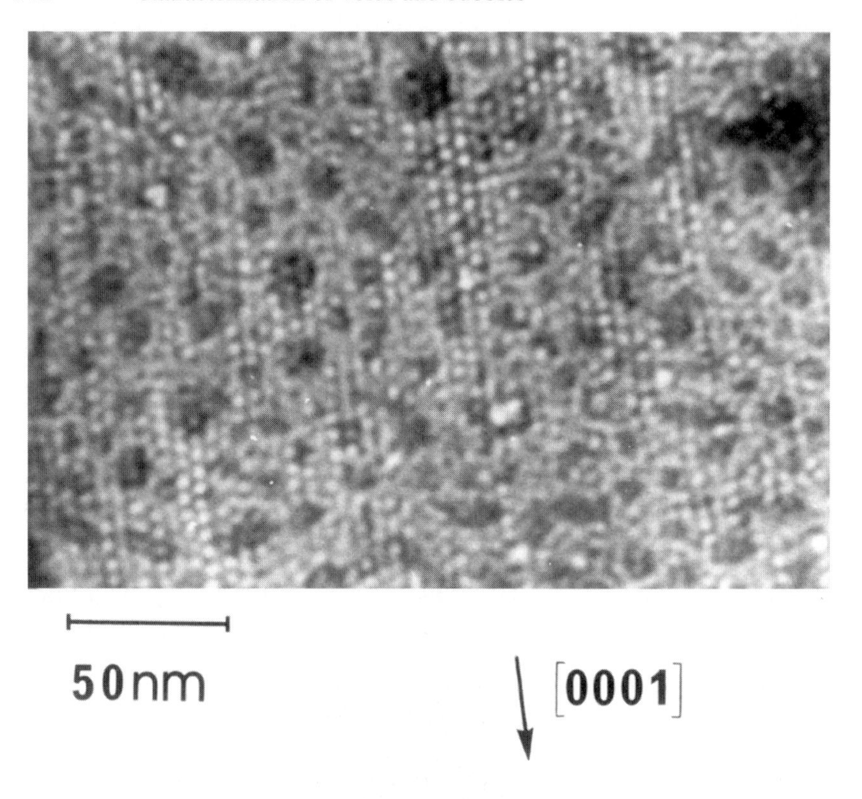

Figure 6.13. Ordering of voids in alumina irradiated to a dose of about 10 dpa with 4 MeV Ar^+ ions at 800 °C. The voids are well-aligned along the *c*-axis and show short-range order in other directions. From Lee *et al* (1985).

dimensional. This phenomenon was first reported for molybdenum irradiated with 2 MeV Ni^+ ions at 870 °C (Evans 1971), but has since been found in several materials after irradiation with ions or neutrons at elevated temperatures. An example of ordering of voids in alumina irradiated to about 10 dpa at 800 °C is shown in figure 6.13. In this case the voids are well-ordered along the *c*-axis, and show short-range order in other directions. An example of a three-dimensional void lattice in pure molybdenum irradiated with fast neutrons to a dose of 2×10^{26} m^{-2} at a temperature of 465 °C is shown in figure 6.14(*a*). The void lattice is identical in structure and parallel to the host lattice. Sass and Eyre (1973) have demonstrated that a periodic array of voids acts as a diffraction grating for electrons, giving rise to extra reflections with spacings inversely related to the void spacings. Although the array is imperfect, its cubic symmetry is evident. The selected-area diffraction pattern of figure 6.14(*b*) shows extra reflections lying along $\langle 011 \rangle$ directions around the (000) beam. These reflections may be indexed as $\{011\}$-type with respect to the void array. The void spacing deduced from these

(*a*)

(*b*)

Figure 6.14. (*a*) A well-developed three-dimensional void lattice in pure molybdenum irradiated with fast neutrons to a dose of 2×10^{26} m^{-2} at a temperature of 465 °C. (*b*) Selected-area diffraction pattern, showing extra reflections around the (000) beam due to the ordered voids. From Sass and Eyre (1973).

extra reflections is 17.1 ± 0.6 nm, in good agreement with direct measurements from the image (18.5 ± 2.5 nm). Diffraction therefore provides an easy method of determining the average void lattice parameter, which is particularly useful if the void or bubble spacing is small. This is the case, for example, in figure 6.15, which shows a fine array of helium bubbles in molybdenum after irradiation with 36 keV alpha-particles at 20 °C. The average spacing of these bubbles is not

Figure 6.15. Zone-refined molybdenum after irradiation with 36 keV alpha-particles at 20 °C to produce helium bubbles: (*a*) bright-field image showing the bubble lattice projected along a [111] direction; (*b*) selected-area diffraction pattern showing the region in the vicinity of the (000) reflection. From Sass and Eyre (1973).

readily measured directly, but is easily found from the diffraction pattern to be 3.7 ± 0.3 nm.

Chapter 7

Techniques for imaging displacement cascades

The majority of displacements in a material subjected to fast neutron irradiation occur in energetic displacement cascades. Displacement cascades have therefore attracted much theoretical attention. In recent years much progress has been made in the modelling of cascades by computer simulation, and a coherent picture of cascade evolution has emerged. Molecular dynamics simulates the displacement and thermal spike phases of the cascade process, which together last less than about 10^{-11} s. In the thermal spike phase a liquid-like phase develops as the energy of the primary knock-on is partitioned among the atoms in the cascade region. Subsequently, this 'molten zone' recrystallizes, leaving a residual number of point-defects which may aggregate to form dislocation loops or other clusters. Clearly, the primary cascade process itself is not amenable to direct experimental investigation. It is possible to study the clusters produced as the end result of the cascade process—for example, the vacancy loops produced by the collapse of vacancy-rich depleted zones at the cascade core. The techniques for doing this have been discussed in previous chapters. Such studies have yielded valuable information on the processes of point-defect survival and aggregation in cascades, but have given only limited information on the cascade process itself. In this chapter we discuss two related techniques which allow more direct 'imaging' of the cascade and comparisons with the predictions of molecular dynamics.

7.1 Imaging disordered zones in ordered alloys

This powerful technique utilizes structure factor contrast to allow the regions covered by displacement cascades in ordered alloys to be revealed. Displacement cascade events in ordered alloys such as Cu_3Au or Ni_3Al lead to the production of local regions of reduced long-range order, so-called 'disordered zones', embedded within the ordered matrix. If certain experimental conditions are met disordered zones can be imaged and directly reveal the cascade structure. Specifically, disordered zones have a reduced structure factor for superlattice reflections compared with the ordered matrix, and so can be imaged in a dark field using superlattice reflections (Jenkins and Wilkens 1976, Jenkins *et al* 1976).

The specific conditions for good imaging of disordered zones in Cu_3Au were determined by contrast calculations and confirmed by experiment. The following results were found:

(i) The contrast of disordered zones in Cu_3Au is a relatively sensitive function of the foil thickness. Figure 7.1 shows the foil thickness dependence of the image contrast of disordered zones in Cu_3Au for superlattice reflections of two types, $\boldsymbol{g} = \{110\}$ and $\boldsymbol{g} = \{100\}$. For $\{110\}$ reflections, the dark-field image intensity is below background ($I/I_0 < 1$) for all values of the foil thickness except for the region $t = 40$–60 nm. For $\{100\}$ reflections the image intensity oscillates both below and above background, suggesting that this reflection is less suitable for imaging disordered zones[1].

(ii) Disordered zones in Cu_3Au are best visible in very thin areas of foil in dark-field images taken with $\boldsymbol{g} = \{110\}$, when they would be expected to appear dark against a light background. For good contrast the foil thickness at an electron energy of 120 kV for a $\{110\}$ reflection must be <35 nm. It may be possible to use somewhat thicker regions at higher operating voltages.

(iii) Provided these conditions are met, disordered zones of size greater than about 2 nm will be visible in dark-field images using a strongly-excited superlattice reflection. The strength of the contrast is proportional to the thickness of the disordered zone in the beam direction, and to the degree of disorder within the zone. The contrast figures may therefore be interpreted as two-dimensional projections of the disordered zones.

(iv) Structure contrast dominates over any strain contrast which may arise from the presence of a dislocation loop produced by cascade collapse, provided that the strain field of the loop is effectively contained within the disordered zone (Wilkens *et al* 1977). Dislocation loops can, however, be imaged by using *fundamental* reflections, which are not sensitive to order.

Figure 7.2 shows disordered zones in Cu_3Au produced by fission neutron irradiation at room temperature, and imaged in a dark field using the superlattice reflection $\boldsymbol{g} = 110$. The disordered zones are visible as dark dots, many of which show considerable fine structure. It is easy to see why such images may be regarded as 'direct' images of cascades. The 15 and 25 nm thickness contours (measured by the convergent-beam technique, see section 5.1) are indicated. Clearly the contrast is best in the region between these contours, in agreement with the contrast calculations of figure 7.1.

The technique has allowed displacement cascade damage produced by various fast particles, including heavy-ions, protons and fast neutrons, to be

[1] In Ni_3Al it is found experimentally that disordered zones show good visibility in thin areas of foil in both $\{100\}$ and $\{110\}$ reflections, although no image simulations have been reported.

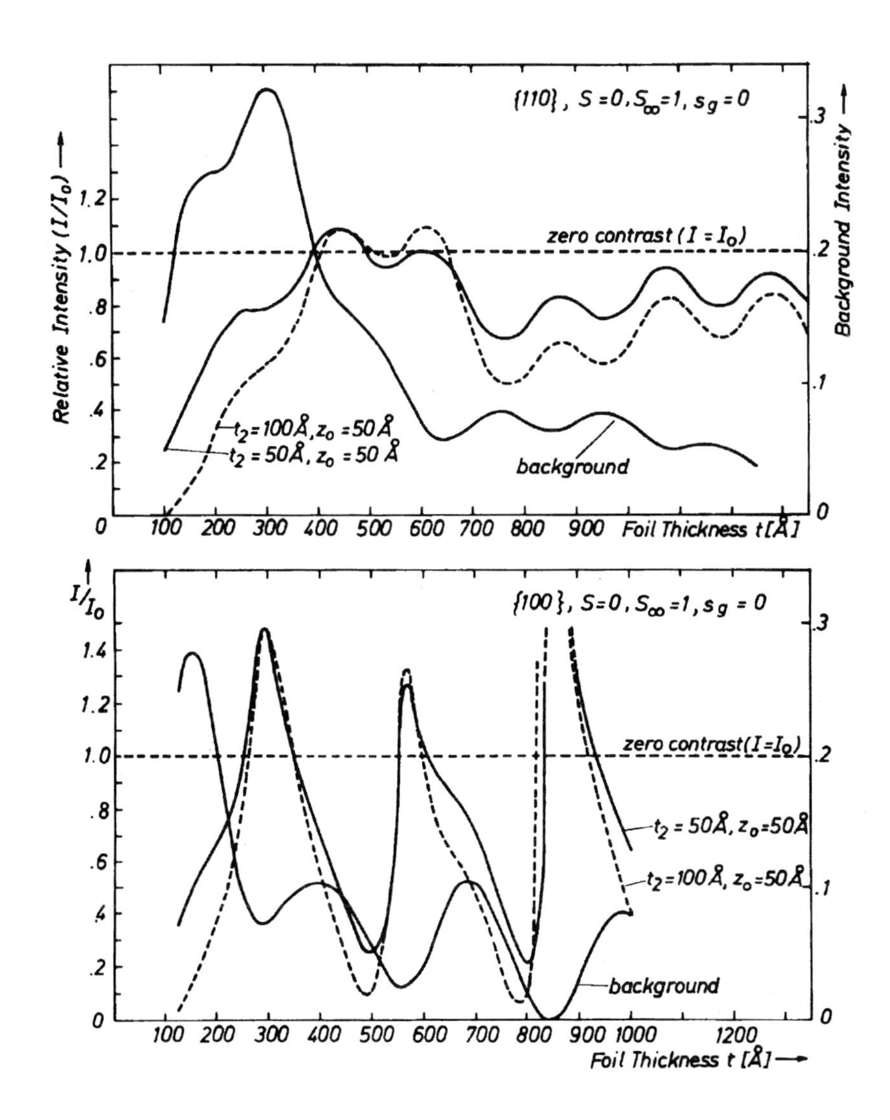

Figure 7.1. The depth dependence of the contrast of disordered zones in (*a*) {110} and (*b*) {100} superlattice dark-field reflections as a function of the foil thickness t. In each case the relative intensity I/I_0 is plotted for two representative values of the thickness t_2 of the disordered zone, $t_2 = 5$ nm and 10 nm. The disordered zone is completely disordered (with Bragg–Williams long-range order parameter $S = 0$) and the matrix is fully ordered ($S = 1$). z_0 is the depth of the centre of the disordered slab in the foil, and in each case $z_0 = 5$ nm. The background intensity I_0 is also shown, for unit incident beam intensity. From Jenkins *et al* (1976).

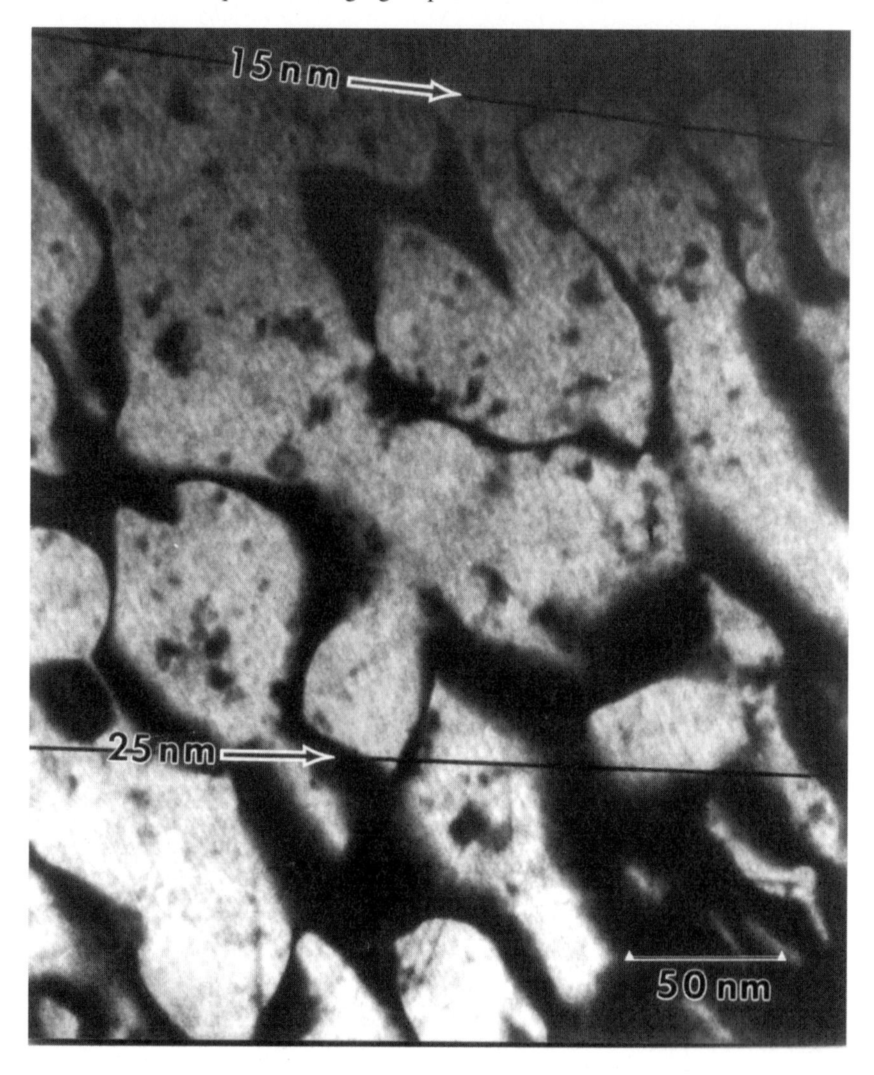

Figure 7.2. Disordered zones in Cu_3Au produced by fission neutron irradiation at room temperature to a dose of 5.3×10^{21} nm^{-2} ($E > 1$ MeV) imaged in the superlattice reflection $\boldsymbol{g} = 110$. The disordered zones, which can be regarded as direct images of the displacement cascades, are visible as dark dots. The 15 nm and 25 nm foil thickness contours are indicated. The larger dark areas are antiphase boundaries.

visualized in several ordered alloys including Cu_3Au, Ni_3Al, Cu_3Pd, Zr_3Al, Ni_3Mn, Fe_3Al, Fe_3Si and Nb_3Sn (see, e.g., Jenkins and English 1982, Howe 1994, Jenkins 1994). With some restrictions imposed by the requirement to use very thin foils, cascade numbers, sizes and shapes have all been quantified. Direct

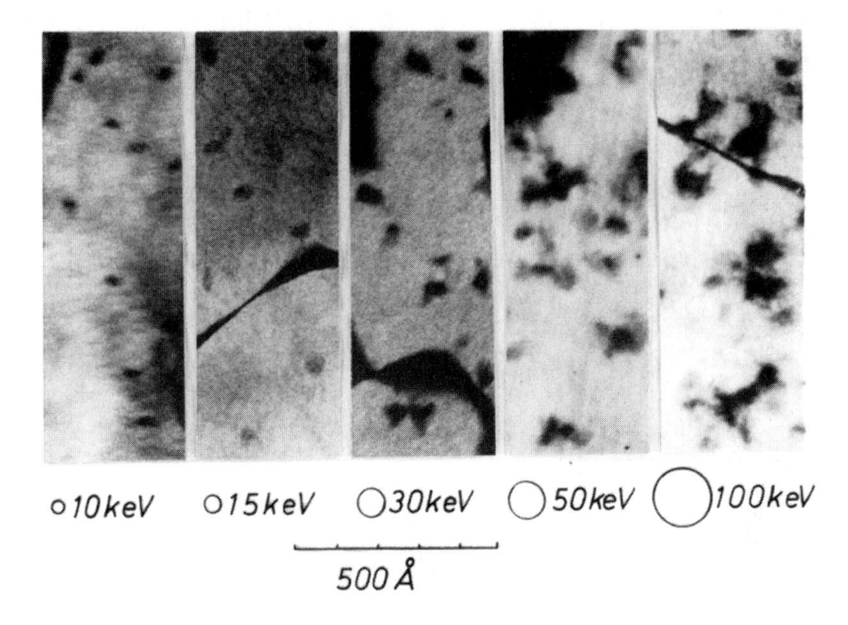

Figure 7.3. Disordered zones in Cu_3Au produced by Cu^+ ion irradiation at room temperature for the various ion energies shown. The circles indicate the expected sizes of displacement cascades D_{th} according to the analytical theory (see next figure). From Jenkins and Wilkens (1976).

comparisons with damage theories and computer simulations have been made. Agreement has generally been good. Some of the more interesting results and conclusions are as follows:

(1) Disordered zones are produced at the majority of cascade sites under most experimental conditions in most of the ordered alloys which have been investigated. For Cu^+ ion irradiations of Cu_3Au at room temperature, all incident ions of energy greater than 10 keV produce a visible disordered zone. Figure 7.3 shows disordered zones in Cu_3Au produced by Cu^+ ions of various energies. The circles indicate the expected sizes of displacement cascades according to theory. It can be seen that the sizes of the disordered zones increase with increasing ion energy in reasonably good agreement with the predictions of theory, particularly at energies below 100 keV. This is shown quantitatively in figure 7.4. The less good agreement at higher energies is partly due to the truncation of cascades by the foil surface, which occurs because of the necessity to use a very thin foil (see Jenkins and English 1982).
(2) Comparisons between experimental and computed images imply that disorder within the cores of the cascades is close to complete. Such

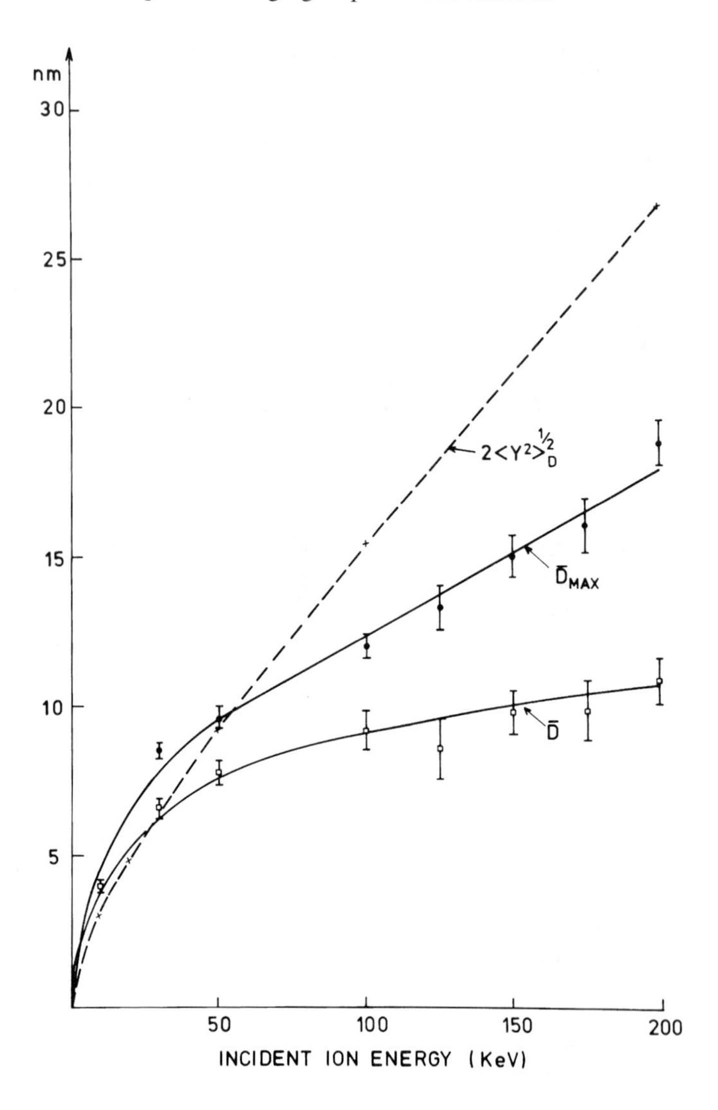

Figure 7.4. Sizes of disordered zones produced by Cu^+ irradiation of Cu_3Au as a function of the incident ion energy. The mean disordered zone diameters $\bar{D}$ and mean maximum disordered zone diameters $\bar{D}_{max}$ are shown, together with the theoretical quantity $D_{th} = 2(\langle Y^2\rangle_D)^{1/2}$ where $\langle Y^2\rangle_D$ is the second moment of the cumulative energy distribution function perpendicular to the incident beam direction. D_{th} gives a theoretical estimate of the lateral extent of cascades at different energies. The trends are in reasonably good agreement with the predictions of the analytical theory in the energy range below about 100 keV. The agreement is less good at higher energies, partly due to thin-foil truncation effects (see Jenkins and English (1982) for a full discussion). From Jenkins *et al* (1979).

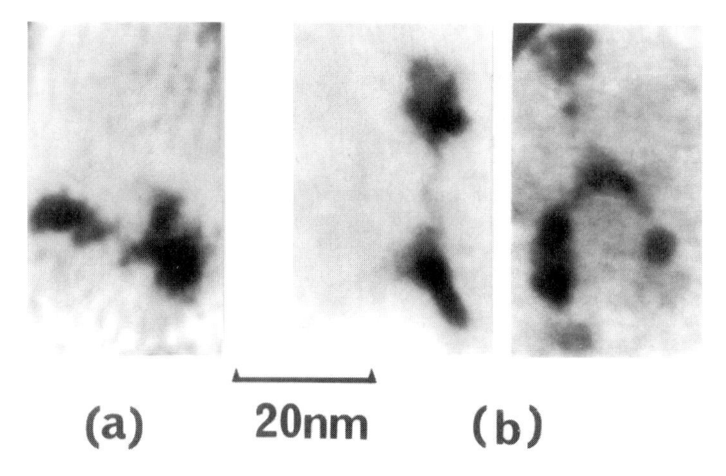

Figure 7.5. Examples of disordered zones in Cu_3Au produced in room-temperature irradiations by (*a*) 150 keV Cu^+ ions and (*b*) 200 keV Cu^+ ions. Subcascade formation is evident. From Jenkins *et al* (1979).

an efficient production of disordered zones at cascade sites is certainly consistent with the idea of local melting, although other mechanisms may contribute to disordering of the cascade peripheries[2].

(3) In higher-energy cascades, disordered zones show increasing fine structure and discrete subcascades develop. Examples for high-energy ion irradiations and fission and fusion neutron irradiations are shown in figures 7.5 and 7.6. Again, these observations are consistent with molecular dynamics simulations, which predict subcascade formation at higher cascade energies. As the ion energy increases, and subcascade formation becomes more marked, the fraction of the theoretical cascade volume occupied by disordered material decreases (see, e.g., Jenkins *et al* 1979, Howe and Rainville 1991).

(4) A particular advantage of the technique is that disordered zones are produced at *all* cascade sites, not just those which collapse to loops. However, those cascades which do contain loops are easily identified by imaging using a fundamental reflection which is sensitive to strain but not to disorder. This makes the technique especially suitable for studying cascade collapse. Absolute measurements of defect yields can be made very simply by taking the ratio of the number of disordered zones which do contain loops to the total number. It is not necessary to know the irradiation dose. Of particular note in experiments of this type are the *in situ* low-temperature experiments of Black *et al* (1987). These authors were able to confirm

[2] For more direct experimental evidence for local melting in cascades in copper see Vetrano *et al* (1990).

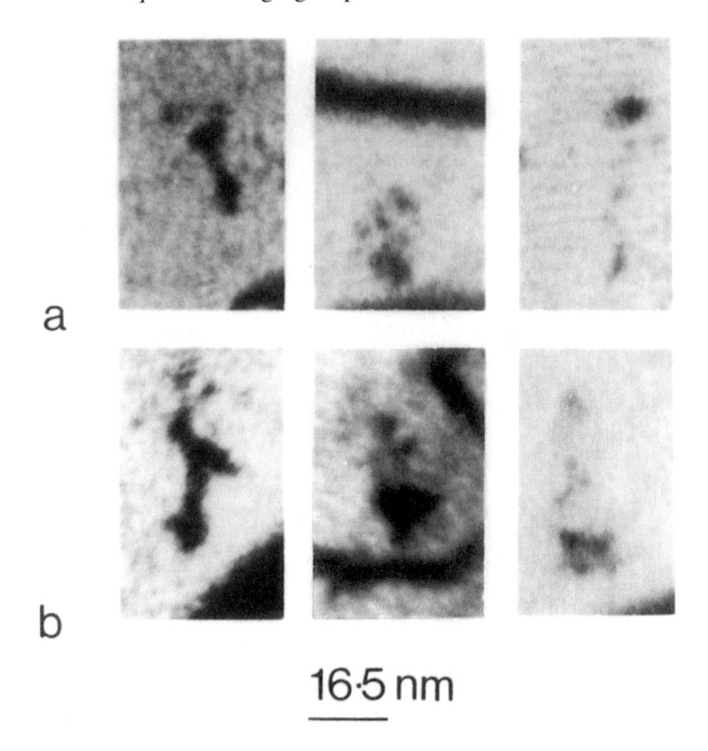

Figure 7.6. Examples of disordered zones in Cu_3Au produced in room-temperature irradiations by (*a*) fission neutrons and (*b*) fusion neutrons. Again, subcascade formation is evident. From English and Jenkins (1981).

directly that cascade collapse in Cu_3Au occurs at an irradiation temperature T_{irr} of 30 K under heavy-ion irradiation. Figure 7.7(*a*) shows an example. These micrographs were recorded at the irradiation temperature of 30 K: note that loops are present at some, but not all, cascade sites. In other experiments, defect yields were measured after warm-up to room temperature (e.g. figure 7.7(*b*)) or after irradiations at room temperature (figure 7.7(*c*)). It was established that collapse to dislocation loops occurred with moderately high probability under all of the irradiation conditions investigated (i.e. irradiations with Ar^+, Cu^+ and Kr^+ ions of energy 50 and 100 keV). There was no subsequent increase in the number of loops on warming to room temperature. However, cascade collapse occurred with significantly higher probability in corresponding room-temperature experiments. These results may be understood within the current model of cascade evolution. At low temperatures, the thermal spike quenches more quickly because of the higher thermal gradients, so that less vacancy aggregation occurs during recrystallization of the molten zone.

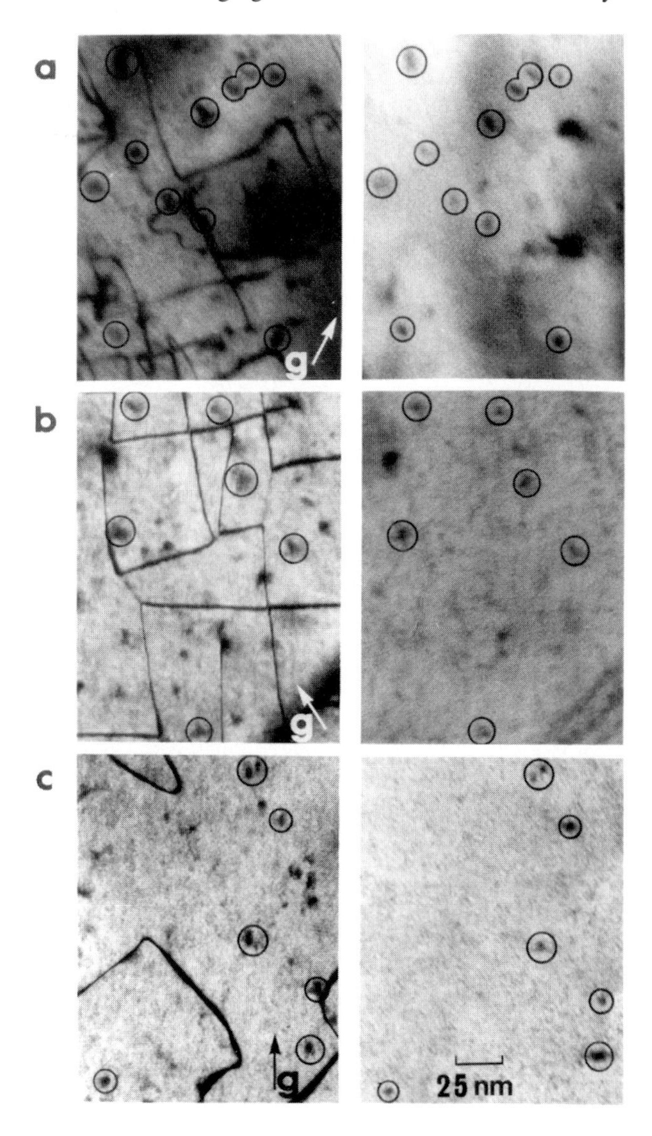

Figure 7.7. Matching of disordered zones with dislocation loops under various irradiation conditions: (*a*) 100 keV Cu^+ at 30 K recorded on the Argonne HVEM at low temperature; (*b*) 50 keV Cu^+ at 30 K, recorded on a conventional TEM at room temperature; and (*c*) 50 keV Ar^+ at 300 K, recorded on a conventional TEM at room temperature. In each case a pair of micrographs of the same area is shown. The left-hand micrograph is an image obtained in a $\boldsymbol{g} = 110$ superlattice reflection, and shows disordered zones (as well as antiphase boundaries). The right-hand micrograph is a bright-field kinematical image using $\boldsymbol{g} = 220$ showing dislocation loops. Some loops and their associated disorded zones are ringed. From Black *et al* (1987).

(5) A similar argument based on thermal spike lifetimes can be used to explain the trends seen in elevated temperature irradiations of Ni_3Al and Cu_3Au. In Ni_3Al irradiated with 300 keV Ni^+ ions, there was a tendency for both normalized number densities (i.e. the number of disordered zones per incident ion) and sizes of disordered zones to decrease with increasing irradiation temperature from 373 to 573 K. A similar tendency was found for 50 keV Ta^+ ions, although the fall-off in number density and sizes occurred at rather higher temperatures (see figure 7.8). In neutron irradiation of Cu_3Au in the Argonne CP-5 reactor, disordered zones produced in an irradiation at 150 °C were smaller and less irregular in shape than disordered zones produced in irradiations at 5 K (Jenkins 1994). Both of these results are consistent with an increased tendency for reordering at the peripheries of disordered zones at higher irradiation temperatures, due to increased lifetimes of thermal spikes. The trends seen in figure 7.8 can be modelled using a simple classical model (Jenkins *et al* 1999b).

7.2 Imaging of amorphous zones in semiconductors, intermetallics and superconductors

A technique related to that described in the previous section uses structure-factor contrast to image amorphous zones at cascade sites in semiconductors, intermetallic compounds and ceramic superconductors. The chief interest in these cases has been the study of the amorphization mechanism itself, rather than fundamental aspects of the cascade process. Nevertheless, some details of the technique are included here for completeness.

Amorphous zones can be imaged in any reflection. An example was shown in figure 2.9. The strength of the contrast depends on the proportion of amorphous material encountered by the electron beam, and so best contrast is obtained in areas of foil on the thin side of the first thickness fringe. An interesting possibility is to use down-zone imaging (section 2.1). The advantage of this over conventional images is shown in figure 7.9, which compares images of amorphous zones in GaAs formed (*a*) in a bright field at the [110] zone axis and (*b*) in a dark field using a strongly excited reflection, $\boldsymbol{g} = 400$. The contrast of the amorphous zones is clearer in the down-axis image, and some zones are seen which do not appear in the dark-field image. In the particular foil thickness in which these micrographs were recorded (stated to be about 26 nm), the contrast of most zones is below background. In the down-zone micrograph, a few have light centres. One of these is arrowed. This contrast effect arises because these zones have a maximum thickness of the order of half the extinction distance under this condition, so that the local foil thickness is effectively reduced by this amount, leading to a reversal in contrast.

The technique has been used to study direct implant amorphization in silicon and GaAs for various implantation species and temperatures (see, e.g., Bench

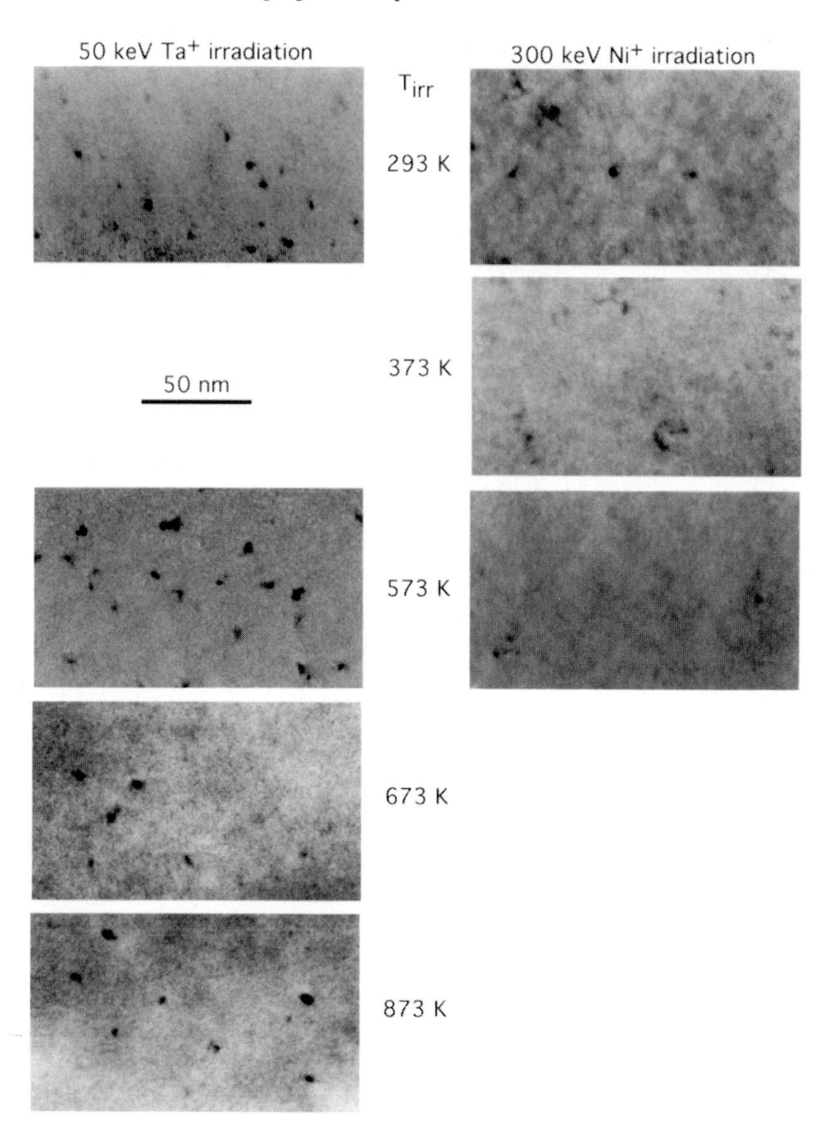

Figure 7.8. Disordered zones in Ni_3Al irradiated with 50 keV Ta^+ ions or 300 keV Ni^+ ions at the various temperatures shown. All images were taken in $\boldsymbol{g} = 100$, and in each case the ion dose was the same. Note the tendency for both the number densities and sizes of disordered zones to decrease with increasing irradiation temperature (see Jenkins *et al* (1999b) for quantitative results).

et al 2000). In GaAs the experiments have to be carried out *in situ*, because amorphous zones anneal with time even at room temperature. Generally, the

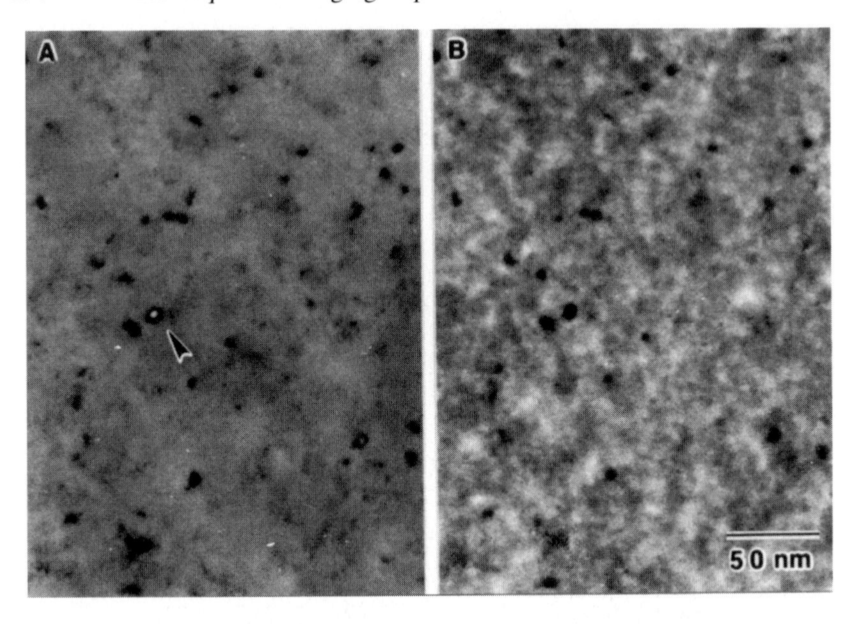

Figure 7.9. Comparison of 'down zone' and conventional dark-field images of amorphous zones in GaAs (after Bench *et al* 1992).

results are consistent with the local-melting model suggested by the computer simulations. Differences in amorphous zone sizes in GaAs between irradiations at 30 and 300 K can again be attributed to a smaller thermal-spike quenching rate at the higher irradiation temperature. At irradiation temperatures above 250 K, recrystallization at the molten zone boundary during the quench leads to smaller amorphous zones. An interesting possibility in GaAs is to image using a (200) superlattice reflection, which is sensitive to anti-site order as was described in section 7.1. In experiments by Bench *et al* (2000), average image sizes in the superlattice reflection (200) were found to be significantly larger than in the fundamental reflection (400), suggesting that the amorphous core region of the cascade is surrounded by a 'disordered' shell rich in anti-site defects.

Annealing of amorphous zones in GaAs can also be stimulated by electron irradiation with electrons of energy less than the displacement energy in crystalline material. It is believed that the process involves excitation and breakage of bonds at the crystalline-to-amorphous interface (Robertson and Jenčič 1996).

Amorphous zones can also be produced directly in cascades in intermetallic compounds and ceramic superconductors. The amorphous nature is often best shown by high-resolution microscopy. An example in the high-temperature superconductor YBCO will be described in section 8.3. An interesting example of cascade damage in the intermetallic compound Zr_3Al has been described

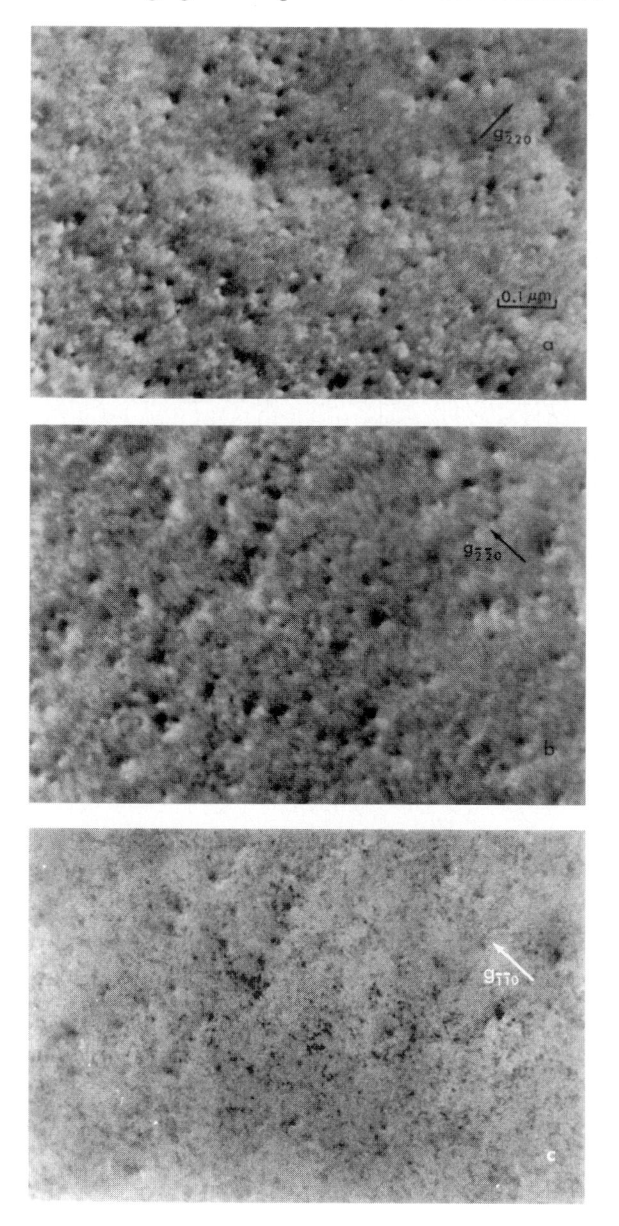

Figure 7.10. Damaged regions produced in Zr_3Al by high-energy (0.5–2 MeV) Ar^+ ion irradiation at 40 K. The dose is 5×10^{15} ion m^{-2} so that individual ion impacts are well separated. Dark-field micrographs taken of the same area under two-beam dynamical conditions using (*a*) the fundamental reflection $\boldsymbol{g} = \bar{2}20$, (*b*) the fundamental reflection $\boldsymbol{g} - \bar{2}\bar{2}0$ and (*c*) the superlattice reflection $\boldsymbol{g} = \bar{1}\bar{1}0$. Zone axis near [001]. From Howe and Rainville (1977).

by Howe and Rainville (1977, 1991). Figure 7.10 depicts damage regions in Zr_3Al produced by high-energy Ar^+ ion irradiation at 40 K. This is a material which can be disordered and then eventually rendered amorphous by irradiation. Figure 7.10(*a*) and (*b*) shows dark-field micrographs of the same field of view taken in the fundamental reflections $\boldsymbol{g} = \bar{2}20$ and $\boldsymbol{g} = \bar{2}\bar{2}0$. Black–white contrast features can be seen with $\boldsymbol{l}$ parallel to $\boldsymbol{g}$ in both micrographs. The contrast is consistent with an outwardly-directed spherically-symmetric strain field. The strain field is dilatational, and so the defects may correspond to clusters of interstitials. However, since amorphous material shows a 5% volume increase compared with crystalline material, the observation is also consistent with the production of amorphous zones at the cores of cascades. Observations of amorphous zones in semiconductors which show the same contrast trends suggest that this second explanation is perhaps more likely. When the same area is imaged using a superlattice reflection (figure 7.10(*c*)), many other smaller damage regions are also visible. These most likely represent disordered zones rather than amorphous regions. Whether amorphization or disordering occurs may be dependent on statistical variations in deposited energy density from cascade to cascade.

Chapter 8

High-resolution imaging of radiation damage

High-resolution imaging has become very important in materials science, particularly in the investigation of interface structures. Its role in radiation-damage studies has been much more limited. To understand why, and to appreciate the potential and limitations of the technique of high-resolution imaging, we need to look briefly at the conditions which must apply for structural imaging.

8.1 Conditions for structural imaging

The basic principles of high-resolution imaging have been described briefly in section 2.3. High-resolution images are obtained by allowing several diffracted beams to contribute to the image. The phase relationships between these beams are optimized by controlled defocusing of the objective lens. The resulting image *may*, under certain conditions, resemble the projected charge density of the object, in which case it is termed a 'structural' image, see figure 2.10. Full details of the theory of high-resolution structural imaging are contained, for example, in the book by Spence (1988).

The conditions for ideal structural imaging can be summarized as follows.

8.1.1 The specimen

(1) There must be a direction in the structure along which columns of heavy atoms line up exactly in projection. In general this requires that the foil should be tilted to a low-index pole orientation. Planar interfaces should be viewed edge-on, if possible with the crystals on both sides of the interface oriented at a low-index pole.
(2) The cell dimensions in the plane of projection should be large enough to allow a good number of diffracted beams to pass through the objective

aperture: the more beams which contribute to the image with similar phases, the better the chance that the image will closely resemble the object. However, the maximum useful angular size of the objective aperture is limited. Usually there is little point in using an aperture of angular size θ larger than that defined by the resolution limits δ_0 or δ_m, i.e. $2\theta < \lambda/\delta$ (see section 8.1.2 below).

(3) The crystal must be very thin—for many specimens <10 nm (although in some structures the same contrast seen in very thin regions may reappear in thicker regions).

(4) Specimens must be sufficiently stable in the electron beam to allow micrographs to be recorded under well-defined conditions. This may require techniques of minimum exposure and image retrieval for beam-sensitive specimens.

8.1.2 The microscope

A dedicated high-resolution electron microscope (HREM) is required for serious high-resolution work, with the following properties:

(5) The resolution must be adequate. Two measures of resolution are commonly applied.

(i) The *point resolution* $\delta_0 = 0.64C_s^{1/4}\lambda^{3/4}$ where C_s is the spherical aberration coefficient of the objective lens (see Spence (1988) for details; δ_0 corresponds to the first zero of the contrast transfer function at the Scherzer defocus). Down to this resolution it is often possible to obtain reliable insight into unknown or aperiodic features such as defects from direct inspection of a *single* micrograph obtained under optimum focus conditions. Below δ_0 no single micrograph can contain all the spatial frequencies and intuitive interpretations are likely to be unreliable.

(ii) The *resolution limit* δ_m ($<\delta_0$, for a good microscope) depends on factors such as the chromatic aberration coefficient C_c, electrical and mechanical stabilities, and beam coherence. δ_m is a measure of the information cut-off. It may be possible to extract information in the resolution range from δ_0 to δ_m by careful comparison of experimental and simulated images for different defocus values.

(6) Other features required of a dedicated HREM include: (a) a high-brightness, high-coherence electron source; (b) a good goniometer stage, allowing the specimen orientation to be adjusted to within 10^{-3} rad (difficult to make compatible with a low C_s lens); (c) controlled defocusing to optimize contrast; and (d) high direct magnifications ($\geq 10^6$) to allow the observation of fine details on the screen. On-line TV systems have now become common, with very high final magnification.

8.2 Image simulations in HREM

If all of these conditions are satisfied, the ideal image of a weak phase object is directly proportional to the projected potential, and so can be interpreted directly and easily. However, real specimens, however thin, are *almost never* good approximations of weak phase objects. For such specimens, dynamical interactions between the diffracted beams and the direct beam must be taken into account. The contrast may then be highly dependent on the specimen thickness and defocus value, and often does not permit direct conclusions to be drawn about the projected crystal structure, or the atomic positions of defects. In such cases it is essential to carry out computer simulations of images, based on some model of the crystal structure or defect, varying both defocus Δf and foil thickness t. The simulated images are then compared with experimental images taken at different defocus values. If differences exist, the model can be modified and a new set of images calculated.

8.3 Applications of HREM to radiation damage

HREM is, in general, *not* well-suited to studies of the detailed atomic structure of radiation-damage defects, because these defects generally lie buried within the matrix and often have complex morphologies. This makes it difficult or impossible to meet the first of the conditions for easy image interpretation described earlier: that there be a projection direction in the structure along which columns of heavy atoms line-up. If this condition is not met, it is generally impossible to achieve a situation where the image closely resembles the projected charge density of the object. Detailed interpretation may not be possible, even with the help of image simulations. In addition, metals and other simple materials, which are usually of most interest in radiation-damage studies, have small unit cells and so may not meet the second and fifth criteria, depending on the quality of the microscope which is available.

Nevertheless if the regions of interest occupy a substantial fraction of the foil thickness it may still be possible to obtain useful images. Usually these are *not* true structural images, and so interpretation requires caution. Some examples where HREM has proven valuable are now described.

8.3.1 Determination of the nature of stacking-fault tetrahedra in silver produced near line dislocations by electron irradiation

The determination of the nature of SFT by high-resolution microscopy was discussed in section 4.4 (see figure 4.15). Essential points are that with a $\langle 110 \rangle$ foil orientation the stacking-faults of most interest lie edge-on in the foil, and the foil thickness is very small (≤ 10 nm). In this case, edge-on stacking faults give interpretable images, which are different for faults of intrinsic or extrinsic character. Sigle *et al* (1988) used this method to confirm that SFT which form near dislocations in electron-irradiated silver (as in figure 2.3) are vacancy in character.

8.3.2 Identification of amorphous and recrystallized zones at cascade sites in the high-temperature superconductor $YBa_2Cu_3O_{7-\delta}$

Heavy-ion irradiation can be used to introduce flux-pinning defects in high-temperature superconductors such as $YBa_2Cu_3O_{7-\delta}$ (YBCO), and so improve the critical current density. We have seen earlier that such defects have a vacancy-type strain field (section 4.2), consistent with amorphous zones. They are most likely produced within 'molten zones' in displacement cascades.

High-resolution imaging does indeed confirm that the defects are amorphous zones. Two defects produced by 50 keV Xe^+ ion irradiation are seen in the dark-field $\boldsymbol{g} = 020$ diffraction contrast image of figure 8.1(*a*). As expected, they show black–white contrast with $\boldsymbol{l}$ parallel to $\boldsymbol{g}$. The high-resolution micrograph of figure 8.1(*b*) shows the same two defects. Clearly, the defects are amorphous zones, but the image also suggests that partial recrystallization has occurred. In this case, the recrystallized region does not seem to be coherent with the matrix (figure 8.1(*c*)).

Another image of a smaller cascade defect in YBCO is shown in figure 8.2. This image is of a defect produced by a 50 keV Kr^+ ion with the irradiation direction near the c-axis. Care should be taken in making size measurements from such images because of the possibility of image delocalization effects. The through-focal-image simulation series of an amorphous region surrounded by a square lattice in figure 8.3 shows how the apparent size of a defect will change with defocus. There is an optimum 'overlap' defocus for any particular spatial frequency u when there is no image delocalization, that is the image of the spacing in question overlaps its true location in the image. Storey *et al* (1996) showed that the overlap defocus is given by

$$\Delta f = -\lambda^2 C_s u^2.$$

The overlap focus for the 0.385 nm fringes seen in figure 8.2 is -160 nm for the conditions of the experiment (electron energy 100 keV, $C_s = 1.75$ mm). This is close to the actual defocus of -140 nm, and so this image should give a good estimate of the true defect size. Note that the best images for sizing are not obtained at the Scherzer defocus of -80 nm. The effect of delocalization may be very strong near Gaussian focus ($\Delta f = 0$), when the structure of the defect may be completely obscured.

Another imaging artefact seen in the image simulations of figure 8.3 is a curving of the lattice planes towards the centre. The magnitude of the effect depends on defocus, and is minimized at the overlap defocus. In an experimental image, fringe bending could easily be misinterpreted as an inwardly directed strain field. This emphasizes the point made earlier that naive interpretations of single high-resolution images may be misleading and should be avoided.

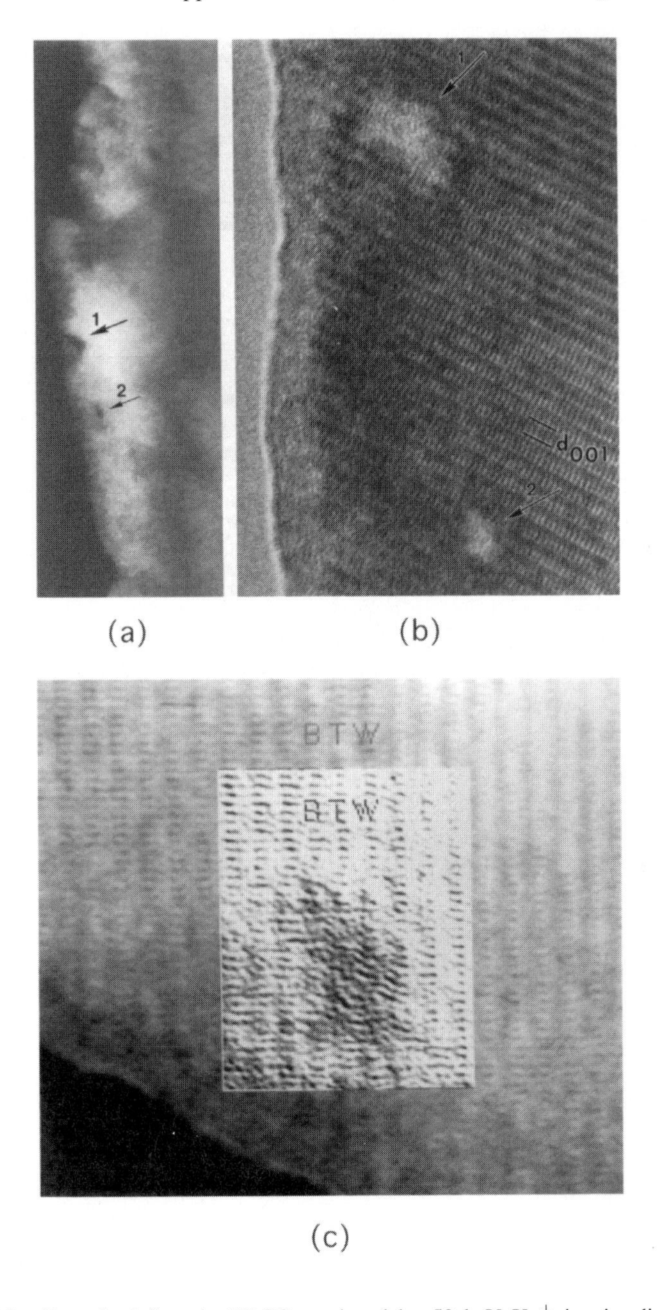

Figure 8.1. Cascade defects in YBCO produced by 50 keV Xe^+ ion irradiation: (*a*) dark-field image in $\boldsymbol{g} = 020$, two defects are arrowed; (*b*) high-resolution image of the same two defects; (*c*) digitized and enhanced version of the image of defect 1 in (*b*). This image is upside-down with respect to (*b*). From Frischherz *et al* (1993).

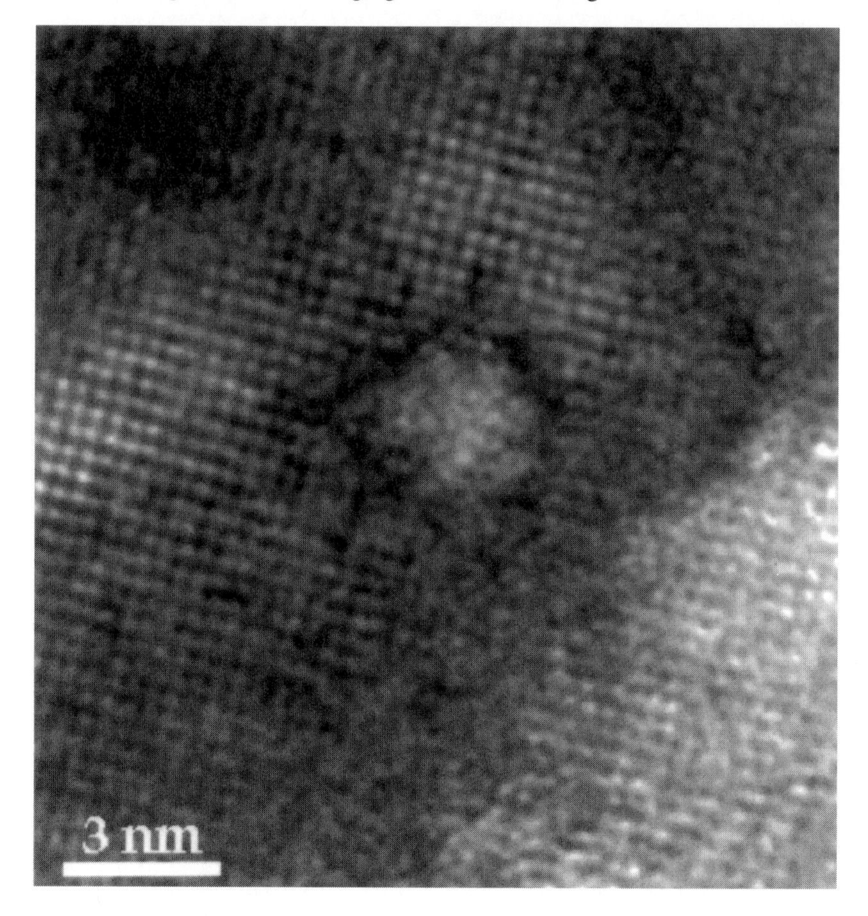

Figure 8.2. Cascade defect in YBCO produced by 50 keV Kr^+ ion irradiation. The objective defocus is about −140 nm, which is close to the 'overlap' defocus of −160 nm, see text. From Storey *et al* (1996).

8.3.3 Identification of the structure of GeV ion tracks, which act as pinning defects in high-temperature superconductors

Another route for introducing flux-pinning defects into YBCO is heavy-ion irradiation at much higher energies, which produces amorphous damage tracks. Such tracks are very effective at pinning magnetic vortices. End-on tracks can be characterized by HREM (Zhu *et al* 1993). The same possibility discussed in the previous section of delocalization artefacts exists.

An interesting feature of ion tracks in YBCO is illustrated in figure 8.4, taken from the work of Yan and Kirk (1998). The structure of fully oxygenated YBCO is orthorhombic due to the ordering of oxygen chains. Specimens typically

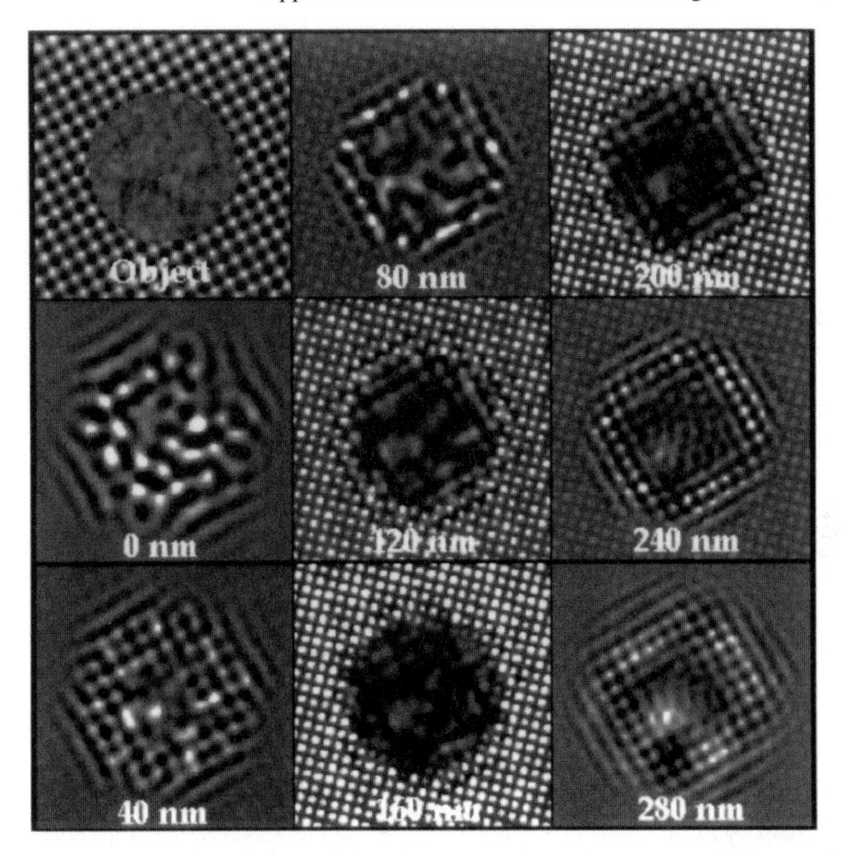

Figure 8.3. Image simulations for the defect model in the top left, which is intended to represent an amorphous cylinder surrounded by a square lattice. The objective defocus is indicated in each image. This series clearly shows two imaging artefacts dependent on the defocus: (1) the apparent size of the cylinder changes; (2) there is an inward bending of lattice planes close to the defect. Both of these effects are minimized at the overlap defocus of -160 nm. From Storey *et al* (1996).

consist of twin-related domains, where the a and b directions interchange. These can be seen in the down-zone bright-field image of figure 8.4(b) as regions of dark and light contrast. The HREM image of figure 8.4(a) shows an individual damage track. In addition to the amorphous core, lobed 'bow-tie' contrast is evident along the b-direction of the matrix. Similar 'bow-tie' contrast is also seen in figure 8.4(b) where it is clear that the contrast lobes show similar contrast to the neighbouring twin domains. Yan and Kirk attribute these observations to 'nano-twinning'. The lobes represent nanometre-sized regions which are in twin orientation with the matrix. Twinning occurs by local re-ordering of the oxygen atoms in the Cu–O chains. The driving force for the local oxygen re-ordering

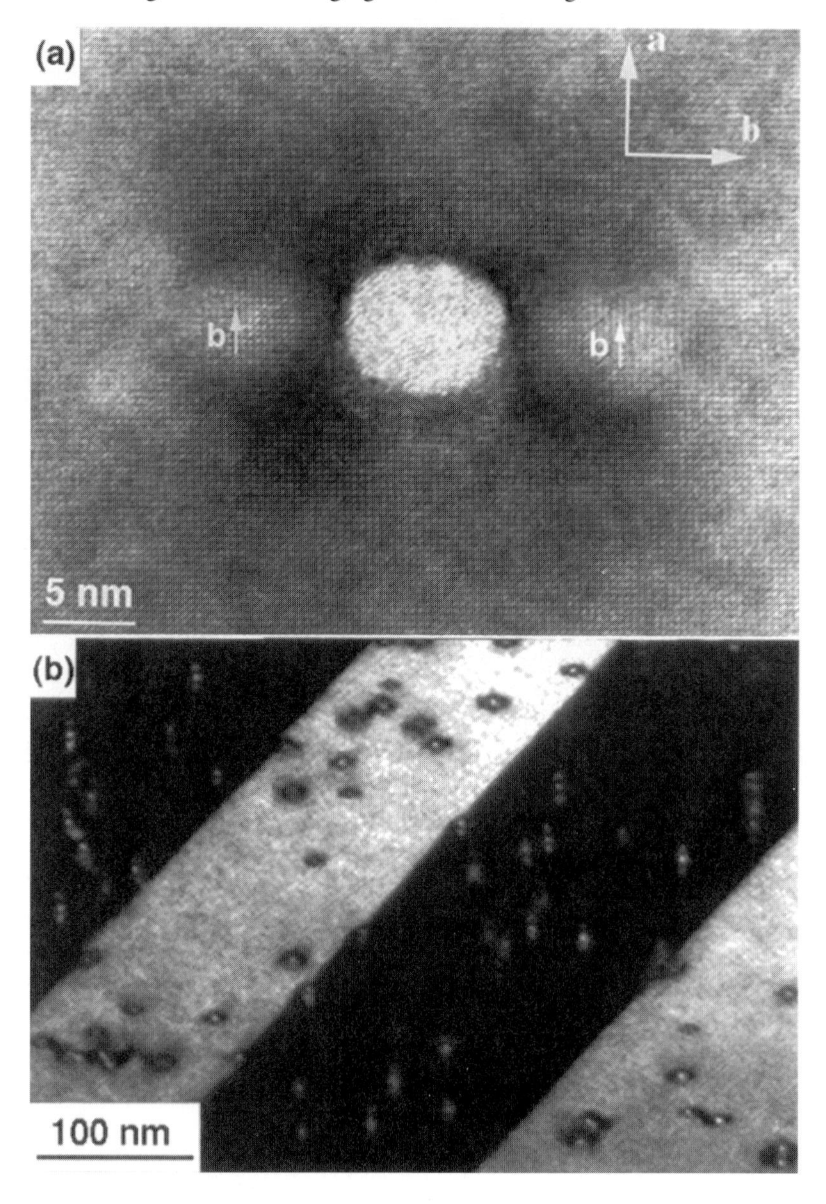

Figure 8.4. Damage tracks in YBCO produced by GeV heavy ions: (*a*) an [001] HREM image of a typical isolated damage track produced by an U ion; (*b*) an [001] zone-axis bright-field image showing damage tracks with several pre-existing twin boundaries in an Au irradiated sample. In (*b*) the specimen was oriented slightly off the exact [001] direction, so that one set of twin domains is in light contrast, and the other is in darker contrast. Note how the 'bow-tie' contrast changes from horizontal in the light domains to vertical in the dark domains. From Yan and Kirk (1998).

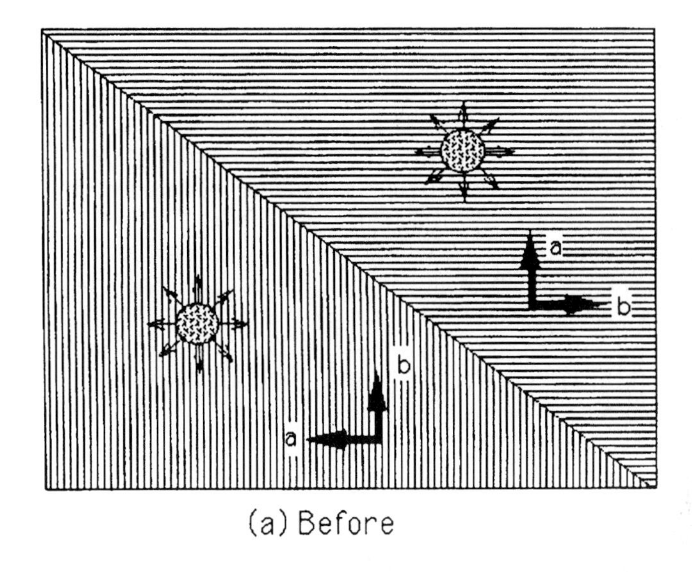

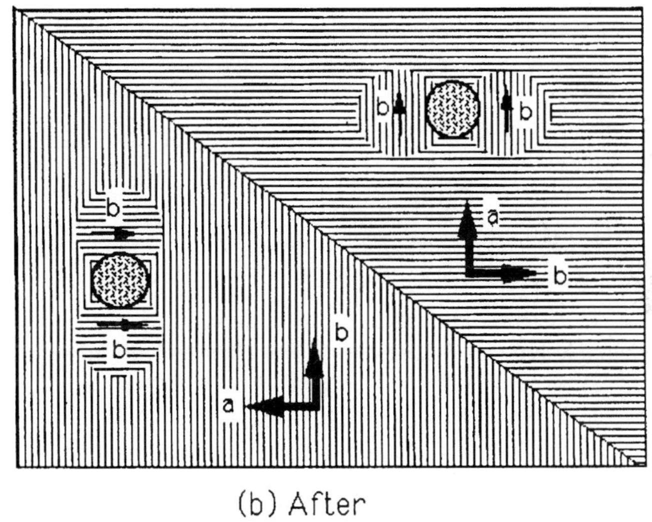

Figure 8.5. Schematic illustration showing how the stress fields can be reduced by local oxygen re-ordering along the *a* and *b* axes of the matrix: (*a*) before; and (*b*) after relaxing the stress. The 'bow-tie' contrast rotates 90° across a pre-existing twin boundary. From Yan and Kirk (1998).

is strain relaxation, shown schematically in figure 8.5. At high ion doses these nanotwins can join up to form a network (Kirk and Yan 1999).

8.3.4 Determination of the structure of copper precipitates in electron and neutron irradiated Fe–Cu alloys

It is well known that neutron-irradiated reactor pressure vessels can be embrittled by the radiation-enhanced formation of copper-rich clusters. The precipitation of copper in ferrite matrices has therefore received much attention, with much of the work focused on thermally-aged model alloys for reasons of simplicity and convenience. Such studies have shown that very small (with diameters of $\sim$1–4 nm) copper-rich precipitates are initially coherent with the ferrite matrix but after reaching a critical size they lose their coherency. HREM studies in aged Fe–Cu and Fe–Cu–Ni model alloys (Othen *et al* 1991, 1994), as well as a martensitic stainless steel (Habibi-Bajguirani and Jenkins 1996), have shown that the copper precipitate structures follow a complicated bcc $\rightarrow$ 9R $\rightarrow$ 3R $\rightarrow$ fcc sequence with increasing ageing time. The 9R structure was determined from HREM images taken at the $[111]_{\mathrm{Fe}}$ orientation. Figure 8.6(*a*) shows a 9R precipitate in a thermally-aged Fe–Cu specimen. The left-hand side of this nearly-spherical precipitate exhibits a characteristic herring-bone fringe pattern which allows the immediate identification of a twinned 9R structure (Othen *et al* 1991, 1994). The fringes have spacings of about three times the expected $(009)_{9\mathrm{R}}$ close-packed plane spacing (i.e. about 0.6 nm). The boundary between adjacent twin bands lies parallel to a set of matrix $\{110\}_{\mathrm{b}}$ fringes. This boundary plane is the $(\bar{1}\bar{1}4)_{9\mathrm{R}}$ plane of mirror symmetry. The orientation relationship between the bcc matrix and the 9R precipitate is therefore $(011)_{\mathrm{b}} \parallel (\bar{1}\bar{1}4)_{9\mathrm{R}}$ and $[1\bar{1}1]_{\mathrm{b}} \parallel [\bar{1}10]_{9\mathrm{R}}$. Note that the characteristic image features which account for the success of HREM in identifying the 9R structure are due to the fact that for this precipitate variant, the $(009)_{9\mathrm{R}}$ close-packed planes and the $(\bar{1}\bar{1}4)_{9\mathrm{R}}$ twin plane are both edge-on to the electron beam at the $[111]_{\mathrm{Fe}}$ orientation.

Some aspects of the complex phase transformations which take place in copper precipitates can be induced by *in situ* electron irradiation with 400 kV electrons. For example, electron irradiation can induce detwinning of 9R precipitates, as shown in figure 8.6(*b*), where it is apparent that one 9R twin variant is growing at the expense of the other. This observation confirms directly the detwinning mechanism proposed by Othen *et al* (1991, 1994). Electron irradiation can also induce rotations of the close-packed $(009)_{9\mathrm{R}}$, which is a first step in the transformation from 9R to 3R (Nicol *et al* 1999). Monzen *et al* (2000) have shown that these plane rotations are associated with a diffusional relaxation of transformation strains. Note that 400 keV is probably below the threshold voltage for knock-on damage in bulk iron and copper at this orientation, but displacements may be occurring at the precipitate–matrix interface.

The same techniques which have been used to investigate copper precipitation under thermal ageing conditions have also been used to study both high-energy electron and neutron irradiation in model Fe–Cu alloys. Hardouin-Duparc *et al* (1996) investigated copper precipitation in an Fe–1.5 wt%Cu alloy irradiated with 2.5 MeV electrons at 295 °C to a dose of 1.4×10^{-3} dpa.

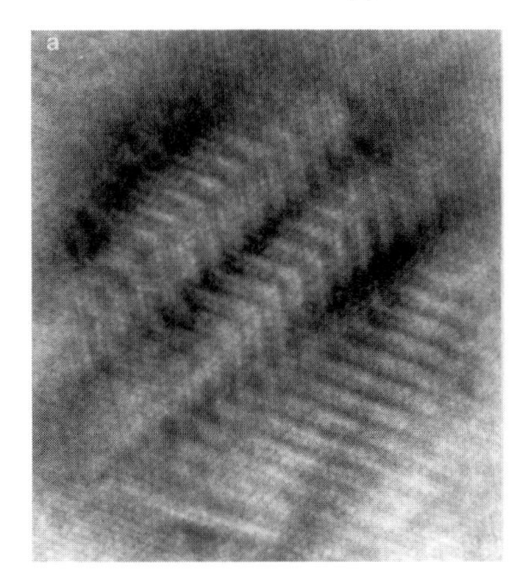

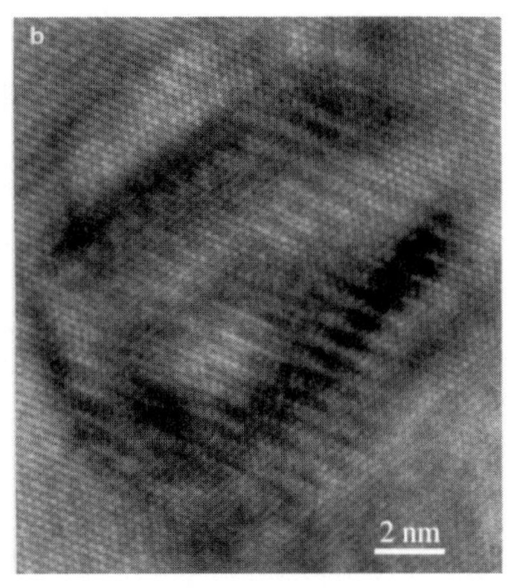

Figure 8.6. *In situ* 400 kV electron irradiation of a twinned 9R precipitate in Fe–1.3wt%Cu in thermally-aged at 550 °C for 10 hr, showing one twin variant growing at the expense of the other. (*a*) At the start of observation, the region to the top left shows the characteristic herring-bone fringes of twinned 9R precipitates, while the region to the bottom right consists of a single twin variant; (*b*) after 10 min under the electron beam, this larger segment has grown through the whole precipitate, which is now untwinned. From Nicol *et al* (1999).

An HREM micrograph of some typical precipitates is shown in figure 8.7. The characteristic herring-bone fringes again confirm that these precipitates have a twinned 9R structure. Hardouin-Duparc *et al* showed that all precipitates in the size range 4–8 nm have a twinned 9R structure. Precipitates with sizes up to about 20 nm were observed and those larger than about 8 nm appeared to have transformed wholly or partially to 3R or fcc.

Nicol *et al* (1999) used HREM to study copper-rich precipitates in an Fe–1.3 wt%Cu model alloy irradiated with neutrons to doses of 8.61×10^{-3} dpa and 6.3×10^{-2} dpa at a temperature of ~270 °C. Again, twinned 9R precipitates of size ~2–4 nm were identified in both materials. Untwinned 9R precipitates in the same size range were also seen. Examples of both types of precipitate in the higher dose material are shown in figure 8.8. The bcc–9R transformation of such small precipitates may be triggered by the passage of a dislocation, a possibility which has also been suggested by molecular dynamics (Harry and Bacon 1997). Such a dislocation-induced transformation process provides an alternative to the generally accepted Russell–Brown hardening mechanism.

8.3.5 Determination of the structure of solid Xe precipitates ('bubbles') in electron-irradiated aluminium

Implanted noble gas atoms are insoluble in metals, and condense to form solid noble-gas precipitates. *In situ* electron irradiation at voltages above the

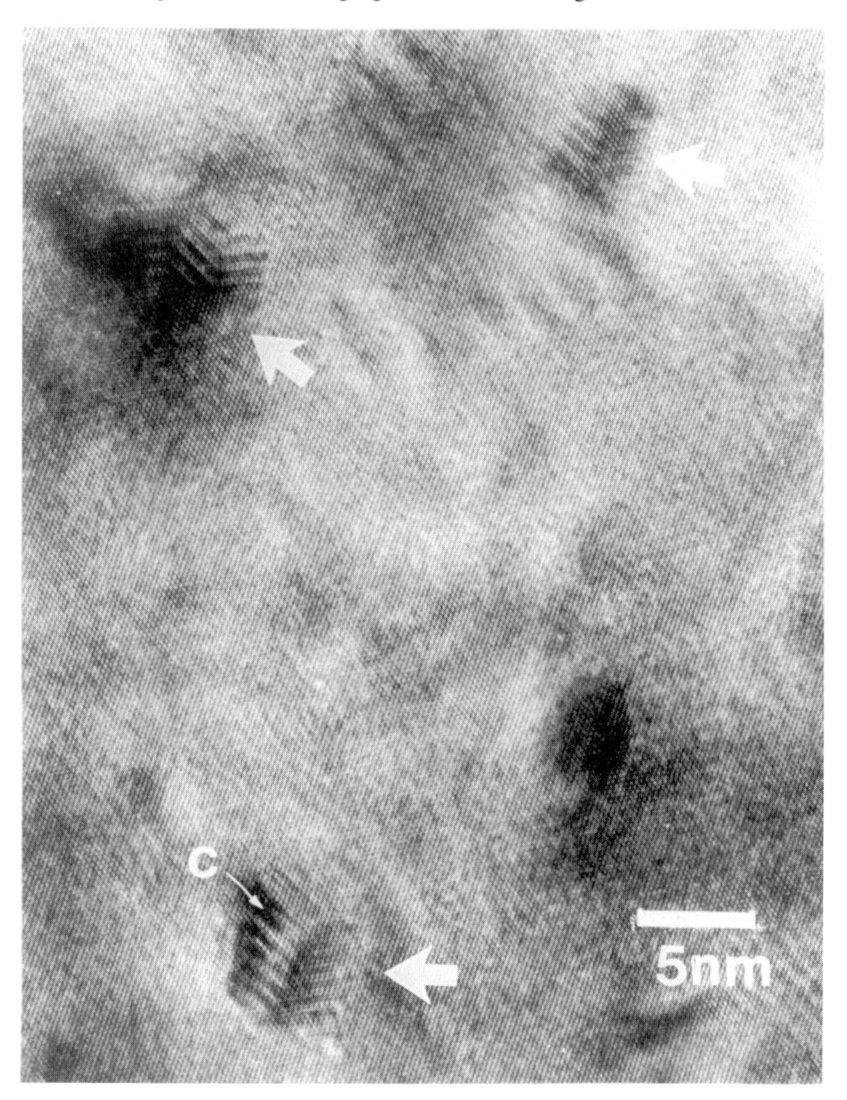

Figure 8.7. High-resolution micrograph taken along $\langle 111 \rangle_{Fe}$ showing as arrowed three small copper precipitates with 9R fringe patterns in an electron-irradiated Fe–Cu alloy. C marks a cubic stacking fault in the precipitate at the bottom of the micrograph. For further details see Hardouin-Duparc *et al* (1996).

threshold for knock-on damage can be used to investigate solid-precipitate growth, migration and coalescence mechanisms (Birtcher *et al* 1999, Allen *et al* 1999).

Two interesting features of the technique used to observe solid Xe

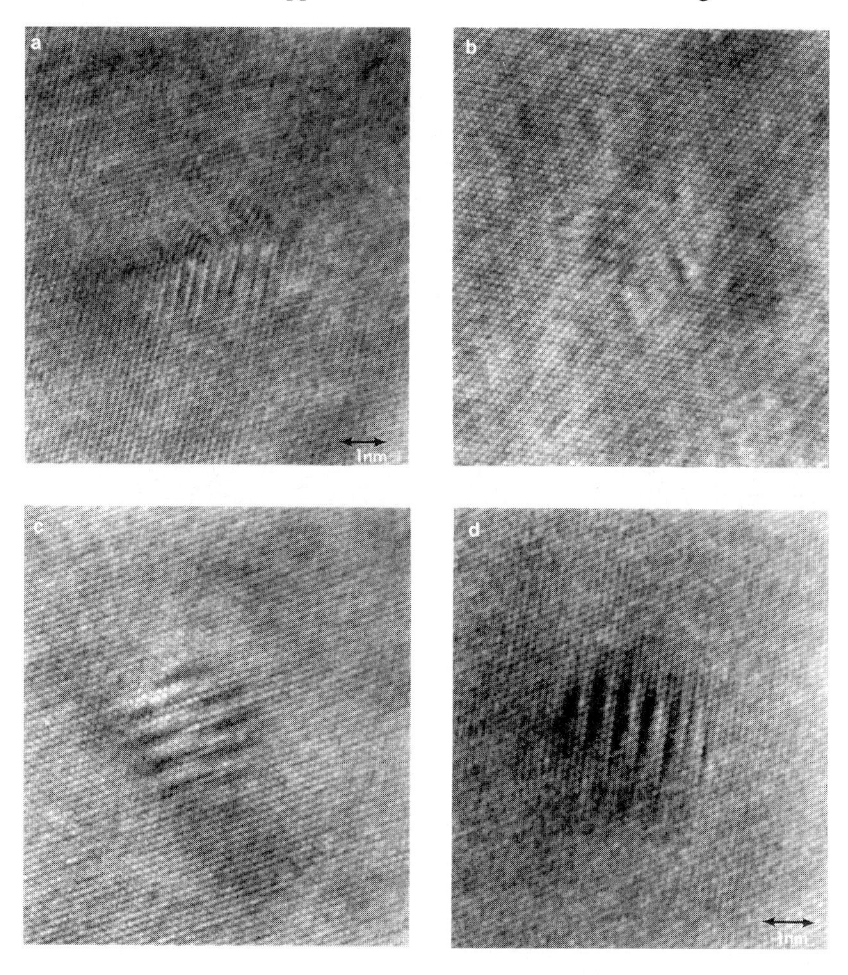

Figure 8.8. High-resolution electron micrographs of copper precipitates in Fe–1.3wt%: neutron-irradiated to a dose of 6.3×10^{-2} dpa at a temperature $\sim$270 °C: (*a*) and (*b*) twinned 9R precipitates; (*c*) and (*d*) untwinned precipitates. Work of Nicol *et al* (1999).

precipitates deserve mention. First, the best images of small solid precipitates are achieved when the foil is tilted a few degrees *away* from the [110] zone axis orientation. Second, the microscope defocus Δf is set at an appropriate *non-Scherzer* value such that spatial periodicities present in the bubble are well transferred but spatial periodicities present in the aluminium matrix are not. These two conditions ensure that contrast from the matrix is suppressed. Essentially the aluminium matrix image is 'detuned'. Solid precipitates containing Xe atom columns with as few as two atoms may be well resolved (Tanaka *et al* 1991, Furuya *et al* 1999). An example demonstrating this has been shown in figure 2.10.

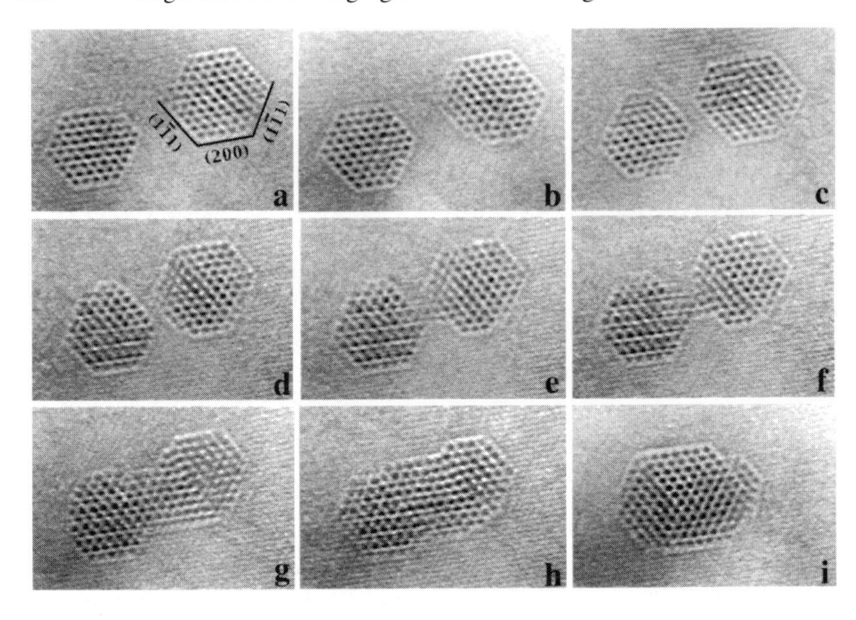

Figure 8.9. Motion and coalescence of two isolated crystalline Xe precipitates during continuous 1 MeV electron irradiation. Measured from the first image (*a*), the elapsed times at which video frames were recorded are: (*b*) 101 s, (*c*) 418 s, (*d*) 549 s, (*e*) 550 s, (*f*) 551 s, (*g*) 561 s, (*h*) 584 s, and (*i*) 727 s. Traces of crystallographic planes are indicated in frame (*a*). From Birtcher *et al* (1999).

Figure 8.9 shows a further example of the coalescence of two solid precipitates under *in situ* 1 MeV electron irradiation at room temperature. The images were obtained from video recordings taken at 30 frames per second, with a running 5 frame average to improve image quality. Shape changes, precipitate motion and eventual coalescence occur due to atomic-level fluctuations of cavity facets induced by the electron beam.

These examples give some idea of the potential of high-resolution imaging. It should, however, be re-emphasized that naive interpretations of HREM images may well be misleading and should be avoided. Comparison with simulated images is always desirable and may be essential.

Chapter 9

In situ irradiation experiments

9.1 Introduction

In situ irradiation experiments in the TEM have played a major role in radiation-damage investigations. The electron beam itself can cause knock-on damage if the electron energy is sufficiently high. The threshold for knock-on damage depends on the material and irradiation direction but is of the order of 400 kV for a medium atomic weight metal such as copper. In high-voltage microscopes operating at 1 MeV displacements are produced in most materials. A displacement rate of 10^{-2} dpa s^{-1} can be achieved, which is about five orders of magnitude higher than in typical reactor neutron irradiations, so high doses can be built up quickly. A dose of say 100 dpa can be accumulated in a single day in an HVEM, which would be equivalent to about 30 years in a power reactor. The use of heating or cooling stages allows good control of the irradiation temperature. It is, therefore, possible to carry out systematic experiments to investigate the nucleation and growth behaviour of point-defect clusters under a wide range of materials and irradiation variables in a fairly short time scale. The damage can be monitored continuously as it evolves. A further advantage is that materials are not activated by the irradiation so that no handling problems arise.

Electron irradiations have the characteristic that point-defects are produced largely as isolated Frenkel pairs. Fast neutrons or high-energy ions produce local regions (10 nm) of atomic damage, often termed cascade damage. Cascade damage can be studied by *in situ* ion irradiations. Ions may be produced accidentally at the electron filament (so-called 'gun-damage') or by coating the filament with an oxide. In some cases, gun damage has been exploited usefully, for example in the study of partially dissociated Frank loops in silver by Jenkins (1974). More often gun damage is undesirable, and most modern instruments have been designed to eliminate it. Several microscope/accelerator facilities have, however, been constructed in order to allow *in situ* ion irradiations under controlled conditions. Typical of these is the Argonne facility, originally based around a 1.2 MV high-voltage microscope (HVEM), but now also including a higher-resolution 300 kV intermediate-voltage TEM (IVEM). These instruments are shown in figure 9.1. Beam lines from ion accelerators are interfaced directly into each microscope. Ions with masses between H and Pb and energies from 20 keV to 4 MeV can be used to irradiate samples. In both microscopes, the sample may be held at controlled temperatures between 12 K and 1000 °C during irradiation and microscopy. Several other similar facilities exist and have been described by Allen *et al* (1994).

All *in situ* irradiations involve thin foils, and this sometimes causes problems in interpretation. The foil surface acts as a strong sink for mobile point-defects so that damage may develop differently than in the bulk material. Provided this is borne in mind, and if necessary taken into account explicitly, *in situ* irradiation experiments can give very powerful insight into damage mechanisms. The potential of such experiments has been enhanced in recent years by the development of much improved video cameras, which have allowed true dynamic experiments to be recorded. Several examples of *in situ* experiments have already been described in earlier chapters. In the following sections, some further examples which illustrate the power of *in situ* irradiation experiments are described.

9.2 *In situ* electron irradiation

Example 1: Measurements of displacement threshold energies and other point-defect properties

HVEM provides a very direct method for measuring displacement threshold energies, simply by determining the minimum electron beam energy which causes the development of radiation damage (often in the form of interstitial dislocation loops) in the area illuminated. Since it is possible to control both the specimen orientation and temperature, the dependence of threshold energies on these parameters can be investigated (Yoshida and Urban 1980, Urban and Yoshida 1981, King *et al* 1983).

We show in figure 9.2 results from King *et al* (1983) on the threshold energy surface of copper measured by *in situ* electron irradiation and resistivity changes at 15 K. The figure shows displacement energy thresholds (in eV) for different specimen orientations in the stereographic triangle. It can be seen that the displacement energy thresholds vary strongly with orientation, being lowest near the $\langle 110 \rangle$ and $\langle 100 \rangle$ poles and highest near the $\langle 111 \rangle$ pole. The difference is about a factor of three. The low values along $\langle 110 \rangle$ and $\langle 100 \rangle$ are probably due to the relative ease of starting replacement collision sequences along these two most close-packed directions. The high value near $\langle 111 \rangle$ arises because this direction is perpendicular to one of the most close-packed $\{111\}$ planes and lies well away from all $\langle 110 \rangle$ and $\langle 100 \rangle$ directions.

Other point-defect properties of metals have been deduced from HVEM experiments. Self-interstitial mobilities can be found from measurements

(*a*)

Figure 9.1. The Argonne HVEM/Tandem Facility: (*a*) the Kratos EM7 high-voltage microscope; (*b*) the newer Hitachi H9000 intermediate-voltage electron microscope. Both microscopes are interfaced to heavy-ion accelerators.

on interstitial cluster nucleation. Vacancy mobilities can be found from measurements on interstitial loop growth rates at high temperatures, or from observations of the shrinkage of interstitial loops, produced at temperatures where

(*b*)

Figure 9.1. (Continued)

vacancies are immobile when the specimen temperature is raised. Kiritani (1994) gives details.

Also using an HVEM, unique experiments have been performed to measure property changes in the high-temperature superconductor $YBa_2Cu_3O_{7-\delta}$ produced by displacement of atoms from specific sublattices in this complex structure. The threshold energy for the formation of a point-defect which pins

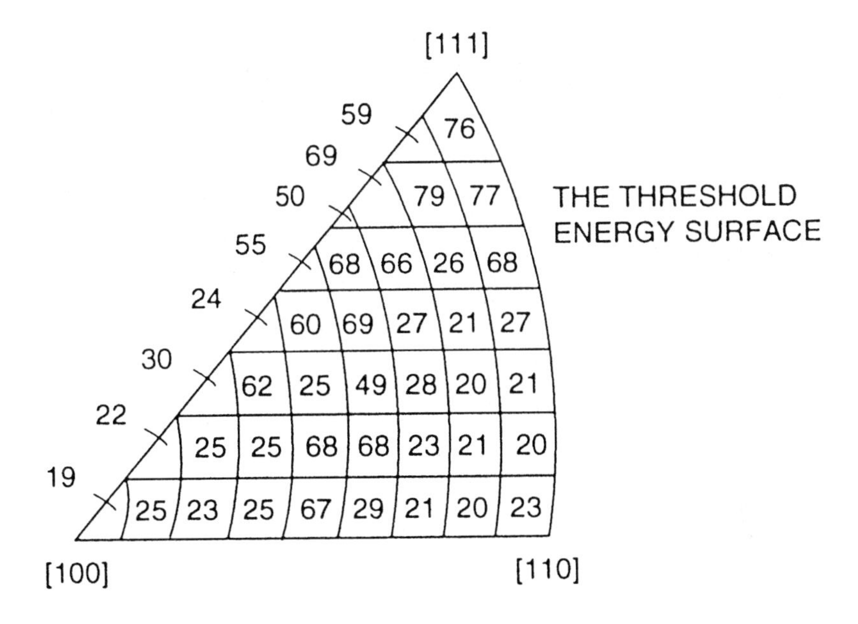

Figure 9.2. Experimental determinations of displacement threshold energies in copper by King *et al* (1983). The displacement energies (in eV) are shown for foils oriented as shown within the stereographic triangle.

magnetic vortices was measured to be that required to displacement Cu atoms (Giapintzakis *et al* 1992). *In situ* low-temperature resistivity was measured to determine the suppression of the critical superconducting temperature by the displacement of oxygen atoms, and from these data the electron pairing state was determined to have d-wave symmetry (Giapintzakis *et al* 1994).

Example 2: Evidence for stochastic fluctuations in point-defect interactions

Electron irradiation is generally supposed to introduce Frenkel pairs homogeneously. Kiritani *et al* (1993), however, have reported observations of the repeated growth and shrinkage of point-defect clusters under HVEM electron irradiation, which are believed to be due to time and space-wise fluctuations in the concentrations of vacancies and interstitials. Figure 9.3 shows a series of micrographs of SFT in 1 MeV electron-irradiated copper, showing repeated cycles of growth by the absorption of vacancies, and shrinkage by the absorption of interstitials. Similar observations have been made in nickel and austenitic stainless steel. Examples of the variations in size of individual SFT with time are shown in figure 9.4. Stochastic fluctuations in point-defect concentrations might be expected to be even more marked in neutron irradiations, where they might have important implications for the nucleation of interstitial loops.

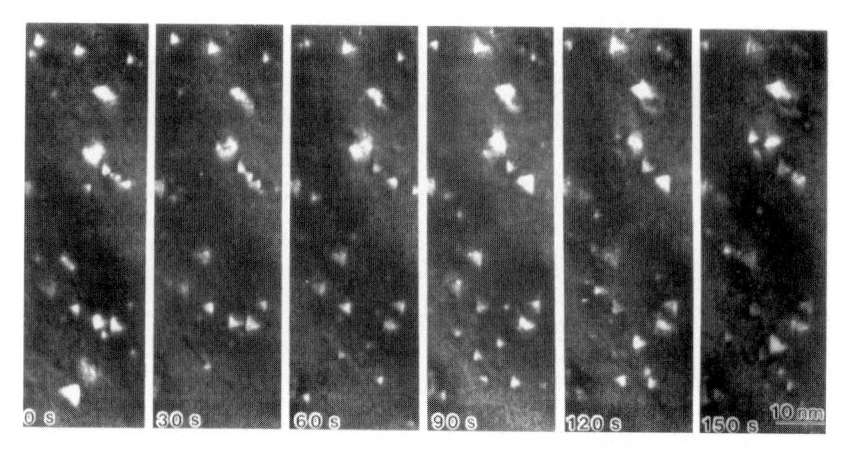

Figure 9.3. Repeated growth and shrinkage of vacancy clusters in copper under 1 MeV electron irradiation at 300 K. The electron flux is 3×10^{23} m^{-2} s^{-1} and the irradiation time is shown. From Kiritani *et al* (1993).

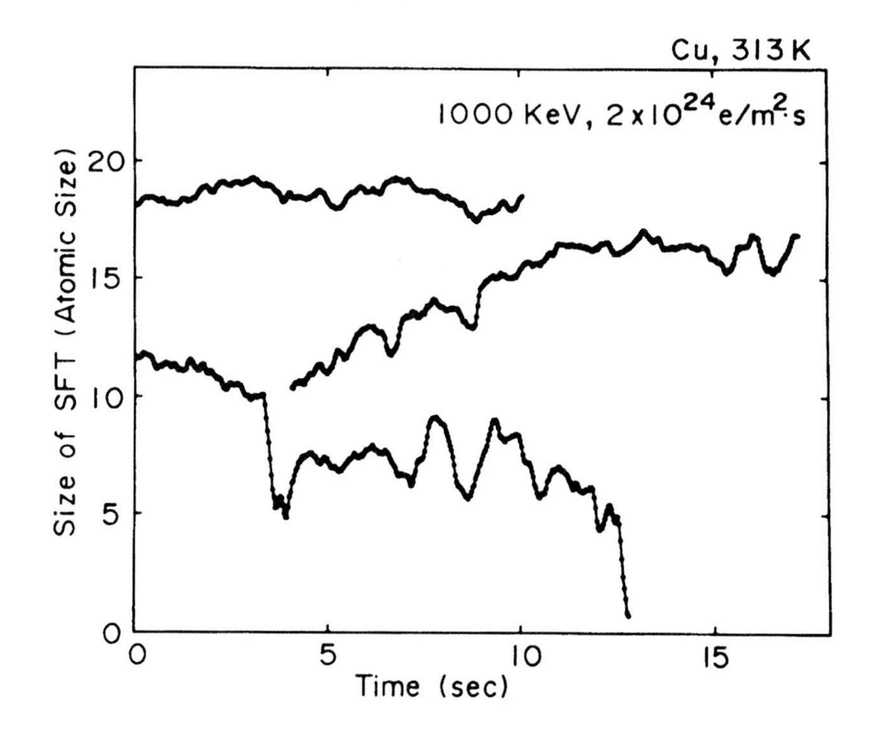

Figure 9.4. Fluctuations in the edge-lengths of individual SFT in copper under 1 MeV electron irradiation at 313 K. From Kiritani *et al* (1993).

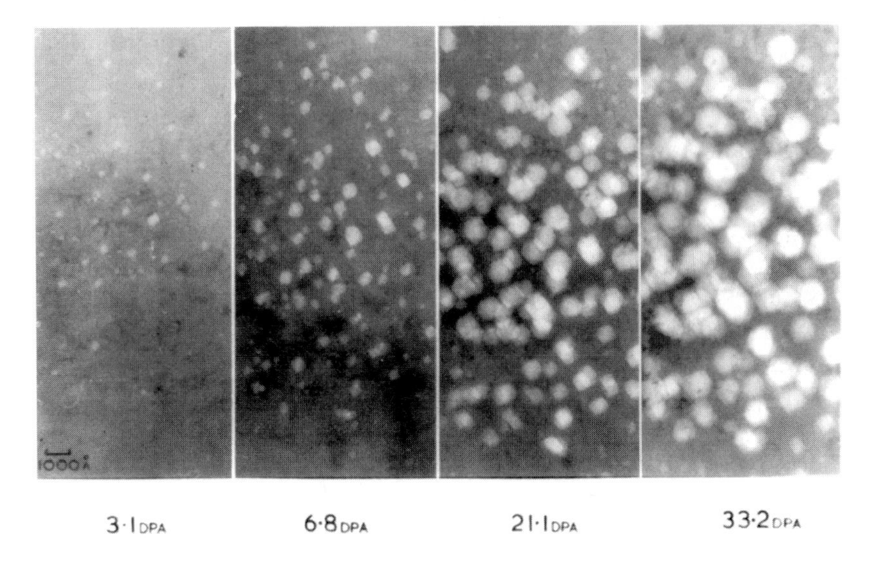

Figure 9.5. Void formation and growth in a Type 316 austenitic stainless steel, under 1 MeV electron irradiation at 600 °C. The figures of displacements per atom (dpa) relate to the number of times, on average, each atom has been displaced from its lattice site by the electron beam. Work of Dr M J Makin.

Example 3: Void swelling

The study of void swelling in stainless steels and other metals was one of the first major applications of HVEM. An example of voids forming and growing in a Type 316 austenitic stainless steel under 1 MeV electron irradiation at 600 °C is shown in figure 9.5. The magnitude of the swelling is a sensitive function of materials parameters as well as irradiation parameters such as temperature and dose. The most important materials parameter is the composition. Variations in major alloying elements affect the type and extent of precipitate phases which occur under thermal treatments and irradiation. Impurities, particularly dissolved gases, have a major affect on void nucleation and stability. The initial mechanical and heat treatment of the alloy is also important. Clearly, separating the influences of these different parameters is a major task. HVEM offers a convenient method for screening potential alloys for use in irradiation environments, so that high-swelling alloys can be identified and excluded from further consideration.

In studying void swelling by electron irradiation the following problems may occur.

(i) The damage is introduced in the form of isolated Frenkel pairs, and so the effects of displacement cascades on void nucleation and growth cannot be studied. The vacancy-rich centres of high-energy cascades in, for example, stainless steels are known to collapse to form Frank loops or SFT even within the void-forming temperature range. This leads to an incubation period before the steady state is achieved, where the loss rate of vacancies due to cascade collapse is balanced by vacancy generation by thermal emission from planar defects. During this incubation period void nucleation and growth is inhibited. The presence of collapsed defects also modifies the sink population, which also has the effect of suppressing void growth.

(ii) Transmutation gases such as hydrogen and helium are produced by neutron irradiation and are known to play an important role in stabilizing void nuclei. Such gases are not introduced by electron irradiation. It is possible to take some account of this by pre-injecting helium into the specimen.

(iii) The steady-state point-defect concentrations are much higher under electron irradiation than under neutron irradiation because of the much higher displacement rates, leading to increased void growth. In general, the relationships between the defect concentrations, defect production rates and other irradiation parameters are nonlinear and quite involved.

(iv) Microstructural development in thin foils might be quite different from the bulk. In particular, the dislocation densities which evolve in thin foils might be quite different from the bulk, leading to different void growth rates and saturation values for the total void volume.

These problems make for uncertainties in extrapolating from accelerated tests to in-reactor performance, and so in-reactor surveillance and test irradiations continue to be necessary. However, there is good evidence that the comparative irradiation response of different materials can be predicted reliably. Williams and Eyre (1976) found that the in-reactor void-swelling characteristics of 316 stainless steel were reasonably well predicted by electron irradiation. Griffiths *et al* (1993) found a good qualitative correlation between the neutron- and electron-damage results in a study of the effects of alloying elements and impurities on radiation-damage development in Zr alloys, although cavity formation occurred more readily under electron irradiation.

Example 4: The crystalline-to-amorphous transition in intermetallics

The response of intermetallic compounds to irradiation is interesting from both a technological and fundamental point of view. Many intermetallics such as Zr_3Al have attractive properties for use in irradiation environments—for example, good mechanical properties at high temperature, good corrosion resistance and low absorption cross sections for thermal neutrons. However, the use of such materials in reactors has been limited since they frequently show irradiation-induced ordered $\rightarrow$ disordered $\rightarrow$ amorphous phase transformations. From a fundamental point of view, irradiation offers a controllable way to destabilize a crystal, and so to study the physics of the crystal-to-glass transition. This is especially the case if amorphization can be induced by electron irradiation. The polymorphous nature of the transformation and the simple homogenous nature of electron damage come closest to matching the assumptions in analytical

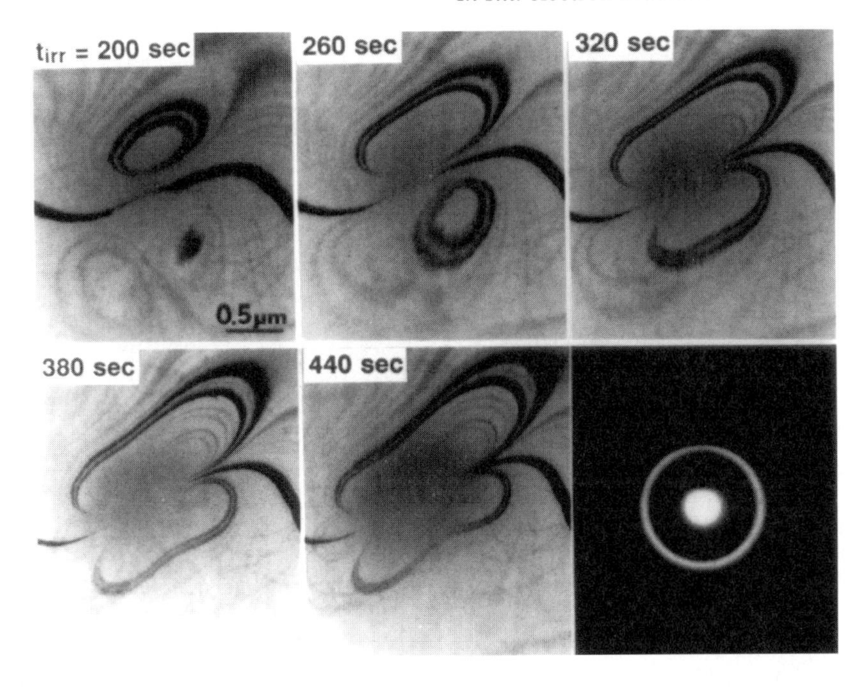

Figure 9.6. A series of bright-field micrographs showing progressive amorphization of CuTi at 10 K with 1 MeV electrons in an HVEM. The irradiation times are indicated. The diffraction pattern at the bottom right was taken from the central region of the micrograph at the end of the experiment and shows complete amorphization. Micrographs from the PhD thesis of Junichi Koike (Northwestern University 1989). For details see also Okamoto *et al* (1999).

theories and modelling of the crystalline to amorphous transformation. The large amount of research in these areas has recently been reviewed by Howe (1994) and Okamoto *et al* (1999). In this section we shall briefly describe *in situ* studies of amorphization of the intermetallic compound CuTi under electron irradiation, which allow the structural changes and kinetics of the process to be observed directly.

Figure 9.6 shows the progressive amorphization of CuTi under electron irradiation at 10 K. Amorphization, which is indicated by the disappearance of crystal-dependent contrast in the image (bend contours) and the development of rings in the diffraction pattern, starts at the centre of the electron beam where the atomic displacement rate is highest. With continued irradiation, the amorphized zone grows radially outward, until the entire area has undergone a homogeneous transformation to the amorphous state. This type of transformation is typical at very low temperatures, when radiation-produced point-defects are immobile. At higher temperatures, the amorphous phase often nucleates preferentially on extended defects such as grain boundaries or dislocations.

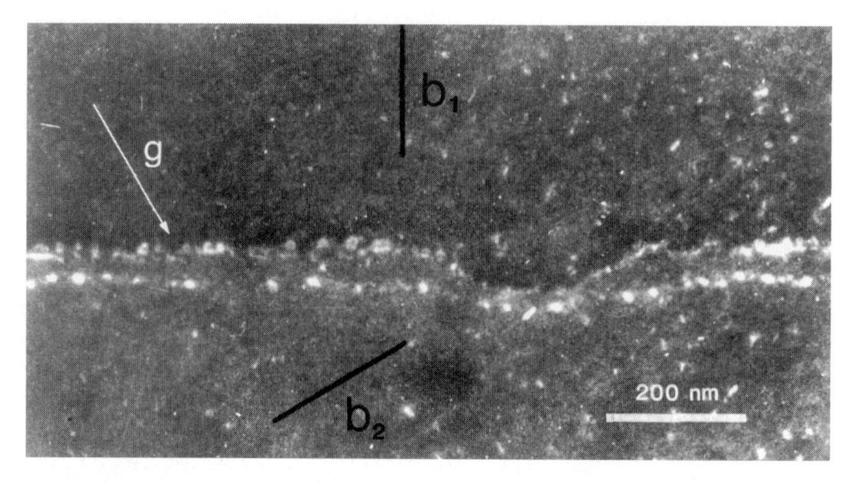

Figure 9.7. Prismatic loops nucleated directly onto individual Shockley partial dislocations in Cu–13%Al after 1 MeV electron irradiation at room temperature. The weak-beam image is taken in a {220} reflection, such that $\boldsymbol{g} \cdot \boldsymbol{b} = 1$ for the top partial dislocation 1, and $\boldsymbol{g} \cdot \boldsymbol{b} = 0$ for the bottom partial dislocation 2 (which is therefore out-of-contrast). The top partial is in the edge orientation over most of its length whilst the bottom partial has an orientation close to 30°. Loops form on both partial dislocations, but more profusely on the top partial dislocation. Micrograph courtesy of Dr D Cherns.

Example 5: Interaction of point-defects with line dislocations in fcc materials

Weak-beam microscopy has been very successful in elucidating the mechanisms of interaction of point-defects with dissociated line dislocations in fcc materials. In low stacking-fault energy Cu–Al alloys with widely dissociated dislocations (among other materials including silicon and GaAs), the classic experiments of Cherns *et al* (1980) established that under electron irradiation, the climb of line dislocations is initiated by the nucleation of prismatic interstitial loops on Shockley partial dislocations with large edge components. An example is shown in figure 9.7. The Burgers vectors of the nucleated loops depend on the dislocation character, and are such as to minimize the elastic energy of the resultant configuration of partial dislocation plus loop, and to maximize the edge component of the loops. Subsequent interactions between these loops and the parent dislocation give rise to very complicated microstructures, which were, however, successfully interpreted by standard weak-beam diffraction contrast analysis. These experiments were performed by electron irradiation at controlled temperatures in an HREM, followed by transfer of the specimen to a conventional TEM for weak-beam imaging.

Here we shall briefly describe a study of the interaction of point-defects with line dislocations in electron-irradiated silver where the experiment was carried out entirely in the Argonne HVEM (Jenkins *et al* 1987). In this case climb does

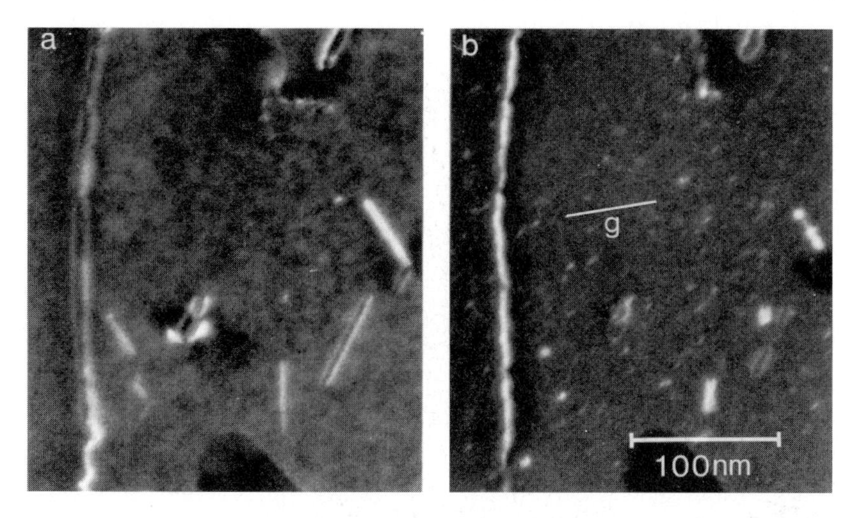

Figure 9.8. An electron-irradiation experiment in silver: (*a*) a dissociated dislocation prior to irradiation, showing well-resolved partial dislocation, $\boldsymbol{g} = 2\bar{2}0$, $|\boldsymbol{g} \cdot \boldsymbol{b}| = 2$; (*b*) after 1 MeV electron irradiation for 1 min, showing the dislocation has constricted, $\boldsymbol{g} = \bar{2}20$, $|\boldsymbol{g} \cdot \boldsymbol{b}| = 2$. From Jenkins *et al* (1987).

not proceed by the mechanism described by Cherns *et al*. Figure 9.8 shows a 75° dislocation in silver lying on a (111) plane. Figure 9.8(*a*) and (*b*) respectively show images before and after 1 MeV electron irradiation to a dose of about 0.05 dpa at room temperature. Each has $|\boldsymbol{g} \cdot \boldsymbol{b}_T| = 2$, where $\boldsymbol{b}_T$ is the total Burgers vector of the dislocation. Before irradiation, figure 9.8(*a*), the dislocation is dissociated into Shockley partials, each producing a well-resolved image peak. The dissociation width is fairly constant and the dislocation is constriction-free. After irradiation, figure 9.8(*b*), the partial separation has narrowed considerably, but the dislocation is still dissociated. No new constrictions have formed, and there is no evidence for major climb motion of the dislocation line. Hirsch (1991) has argued that this narrowing is due to relaxation of the dislocation strain fields by an accumulation of point-defects at the dislocation, interstitials to the tensile side of the dislocation and vacancies to the compressive side. With continued electron irradiation, defect clusters become visible close to the line. In figure 9.9, the fine cluster damage near the dislocation shows the characteristic V-shaped contrast of SFT particulary after the longer irradiation time. Note that in these micrographs $\boldsymbol{g} \cdot \boldsymbol{b}_T = 0$ so the dislocation itself is not seen. Examples of individual SFT which grow with increasing irradiation time are arrowed. These tetrahedra have been shown to be of vacancy type by structural imaging (see section 4.4 and figure 4.15). Similar observations have been made in Fe–Ni–Cr alloys electron-irradiated at 400 °C (King *et al* 1993).

This and other experiments suggest that the mechanism of interaction of

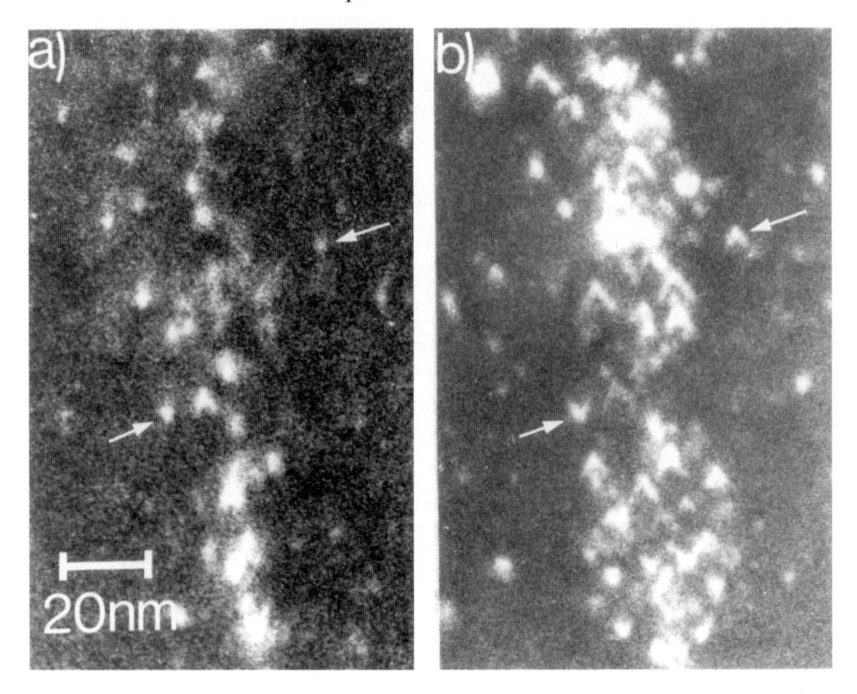

Figure 9.9. The formation of point-defect clusters near the dislocation seen in figure 9.8 with continued 1 MeV electron irradiation. Weak-beam micrographs taken at the [112] pole with (*a*) $\boldsymbol{g} = \bar{1}\bar{1}1$ after 5 min, and (*b*) $\boldsymbol{g} = 11\bar{1}$ after 13 min total electron irradiation time. The dislocation has $|\boldsymbol{g} \cdot \boldsymbol{b}| = 0$ and is out of contrast. The defects seen in (*a*) have grown to be recognizable SFT in (*b*). From Jenkins *et al* (1987).

interstitials with dissociated dislocations is controlled by the relative ease of jog nucleation. In materials where jog nucleation is easy, the interstitial concentration at the dislocation core builds up to a point where interstitial loops nucleate, and the Cherns *et al* climb mechanism occurs. In materials where jog nucleation is difficult, interstitials move along the almost-constricted dislocation line by pipe diffusion to existing jogs or the surface. In thin-foil experiments, interstitial loss to the foil surface will be particularly important. Vacancy supersaturations in regions of compressional strain near dislocations can then increase to the point where vacancy clusters such as SFT can nucleate.

9.3 *In situ* ion irradiation

Example 1: Cascade collapse at low temperature

Vacancy loops are known to be produced in many metals and alloys by the 'collapse' of depleted zones at the cores of displacement cascades. An important question is whether this collapse process is temperature dependent. This question has been addressed by carrying out *in situ* ion irradiations at both low and high irradiation temperatures. In this section we review some of the low-temperature experiments.

Perhaps the most convincing low-temperature experiment was carried out by Black *et al* (1987) in Cu_3Au. In this ordered alloy, 'disordered zones' are produced at the sites of displacement cascades, which can be imaged using dark-field superlattice reflections. Those cascades which collapse to vacancy loops can be identified by imaging the loops in strain contrast using a fundamental reflection. This experiment was described more fully in section 7.1, figure 7.7. Other experiments have been carried out on a number of pure metals including copper, nickel and iron. In all these cases vacancy loops were found at 30 K after heavy-ion irradiation at the same temperature. Since at this temperature no thermal migration of vacancies is possible, this observation indicates that collapse may take place athermally, and probably occurs in the thermal spike phase of the cascade. The collapse probabilities, however, were considerably smaller at 30 K than at room temperature, indicating a thermal component to the collapse process.

Results from the experiments of Robertson *et al* (1991) in nickel are shown in figures 9.10 and 9.11. These authors studied cascade collapse in nickel following irradiation with 50 keV and 100 keV Ni^+ ions and 50 keV Kr^+ ions in the HVEM accelerator facility at Argonne National Laboratory. Vacancy loop formation was followed as a function of both irradiation temperature (30 K and room temperature) and dose (10^{15}–10^{17} ion m^{-2}). Dislocation loops were produced in the low-temperature irradiations, but the probability of collapse at 30 K was only about half of the room-temperature value. However, a new population of loops appeared when foils irradiated at 30 K were subsequently warmed to room temperature (see figure 9.10). The nature of these new loops was not established directly. The assumption made by Robertson *et al* was that they were of vacancy character, although recent experiments in copper throw some doubt on this (see section 5.2.1). The appearance of loops on warming was attributed to a population of submicroscopic clusters which grow to a visible size when there is at least limited point-defect mobility.

At cascade overlap doses, Robertson *et al* (1991) and Vetrano *et al* (1992) found a new phenomenon of 'cascade dissolution'. With successive incremental doses a fraction of the loops disappeared. Others changed their positions or their Burgers vectors or appeared to coalesce with neighbouring loops (see figure 9.11). These effects were believed to originate from existing loops being engulfed by new cascades. This explanation is given credence by the molecular dynamics simulations of English *et al* (1990) which show that pre-existing loops may be annihilated by a new cascade event and that, in some cases, a new loop may reform.

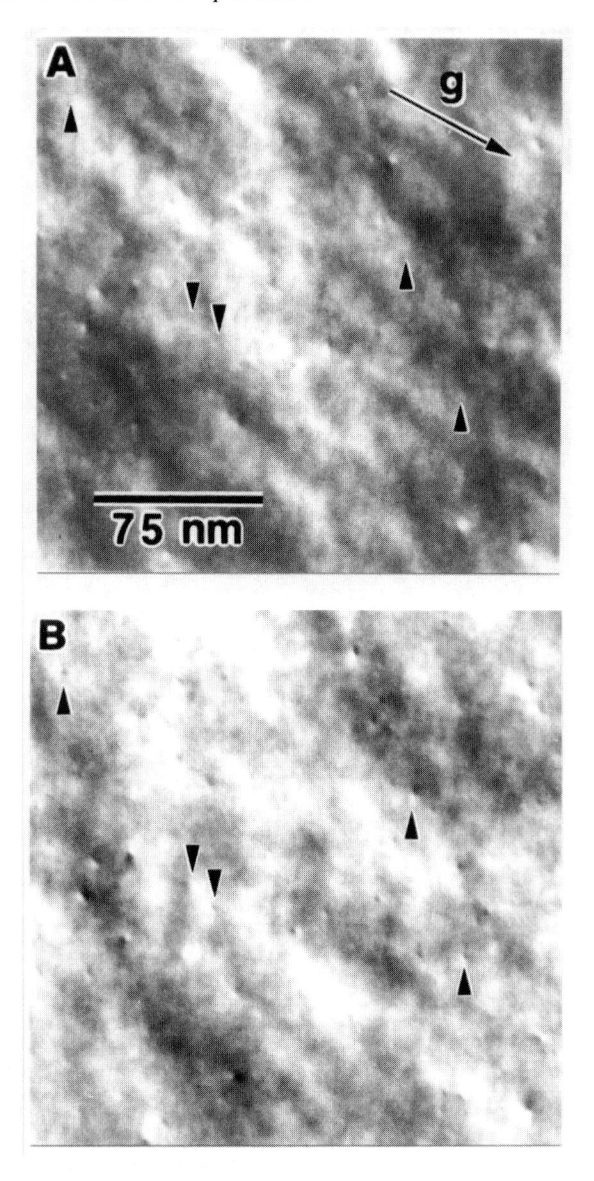

Figure 9.10. Dislocation loops in the same area of a specimen of nickel after irradiation with 50 keV Kr^+ ions at 30 K to a dose of 1.2×10^{16} ion m^{-2}: (*a*) with the specimen still at 30 K and (*b*) after the specimen had warmed to room temperature. The arrowheads mark defects which have appeared on warm-up (*b*) and their location in (*a*). Grain orientation [110], $\boldsymbol{g} = 002$. From Robertson *et al* (1991).

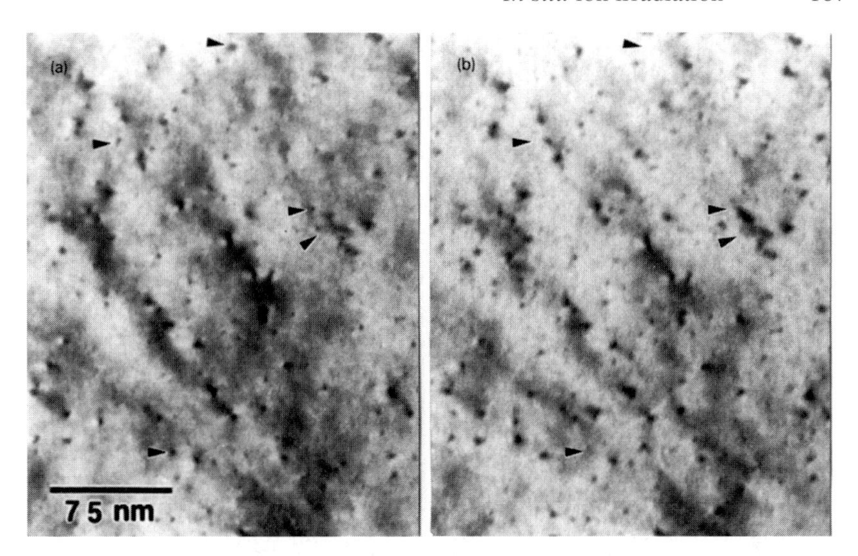

Figure 9.11. Micrographs comparing the same area of a nickel specimen after room-temperature irradiation with 100 keV Ni^+ ions to ion doses of (*a*) 3.4×10^{16}, and (*b*) 4.4×10^{16} ion m^{-2}. The defects marked by arrowheads are present in (*a*) but not in (*b*). Grain orientation [100], $\mathbf{g} = 002$. From Robertson *et al* (1991).

Example 2: Cascade collapse at elevated temperatures

Reactor components normally operate at elevated temperatures. For example, the pressure vessel of a pressurized water reactor is normally at a temperature of about 550 K, whilst near-core components of fast reactors may be at a temperature of 800 K or more. The question of whether cascade collapse occurs with the same probability at elevated temperatures as at room temperature is therefore an important one.

Early experiments using *ex situ* ion irradiations presented a rather mixed picture. In copper the defect yield (that is, the number of visible dislocation loops per incident ion) did fall off with increasing irradiation temperature (English *et al* 1976). In copper irradiated with 30 keV Cu^+ ions, the yield remained constant to a temperature of 573 K but there was a rapid fall-off between 573 and 673 K. English *et al* argued, however, against a true temperature dependence of the defect yield in copper. Rather, they attributed the apparent variation in yield with temperature to thermal emission of vacancies from vacancy loops during the 60 s that the specimen was at elevated temperature during and following the irradiation. A similar argument could explain the drop in yield found in 80 keV W^+-irradiated nickel (Robinson and Jenkins 1981).

Further *ex situ* experiments in molybdenum could not be explained in this way (English *et al* 1977). In specimens irradiated with 60 keV self-ions the number of visible dislocation loops fell systematically over the whole temperature range studied (293–810 K). This was considered to rule out any model involving cascade collapse to loops, followed by thermal vacancy emission, since in this case the vacancy loop population would be expected to remain constant to temperatures >1000 K. English *et al* (1977) therefore concluded that in molybdenum the process of cascade collapse to loops is strongly temperature dependent.

In situ heavy-ion irradiation experiments were pioneered by the group led by Ishino and Sekimura at the University of Tokyo. This group has studied cascade damage in gold, nickel and copper at elevated temperatures, and found that defect clusters are produced 'instantaneously', consistent with their formation by a process of cascade collapse (see, for example, Ishino *et al* 1986). Vacancy loops, SFT and unresolved contrast dots could all be seen to be formed in this manner. The subsequent shrinkage of these defects, both under irradiation and thermal annealing conditions, could also be monitored.

Recently, Daulton *et al* (2000) have used *in situ* microscopy to re-examine quantitatively cascade collapse at elevated temperatures in copper irradiated with 100 keV Kr^+ ions. It was possible to measure 'post-irradiation' yields approximately ± 0.1 s following completion of the ion irradiation from an analysis of video-cassette images. The post-irradiation yield is the number of visible clusters per incident ion, immediately after the ion irradiation, and so is a true measure of the cascade collapse probability at elevated temperatures. These *in situ* experiments also allowed the thermal stability of loops to be investigated experimentally. It was therefore possible to determine post-irradiation yields from conventional micrographs recorded on film some time after the ion irradiation by taking into account the isothermal anneal losses which would have occurred in the interim. The results of these two independent methods of measuring the post-irradiation yields are shown in figure 9.12. They are in excellent agreement. The yield remains approximately constant up to a temperature of 573 K, and then abruptly decreases. This is a true drop in yield, and is not caused by defect loss during or following ion irradiation. This experiment therefore shows unequivocally that the defect yield in heavy-ion irradiations of copper is temperature dependent. The variation of the yield with temperature must result from a temperature dependence of the probability of cascade collapse into defect clusters. This has important consequences for understanding damage development under cascade-producing irradiation at elevated temperatures.

Example 3: Formation of craters and holes by single ion impacts

That near-surface displacement cascades in dense materials such as gold and molybdenum can produce surface craters has been recognized for many years (e.g. Merkle and Jäger 1981, English and Jenkins 1987). An example of craters produced in molybdenum by heavy-ion irradiation was shown in figure 6.3. In thin films single cascade events can even lead to the formation of holes. Recent *in situ* experiments by Birtcher and Donnelley (1996, 1998) have revealed new

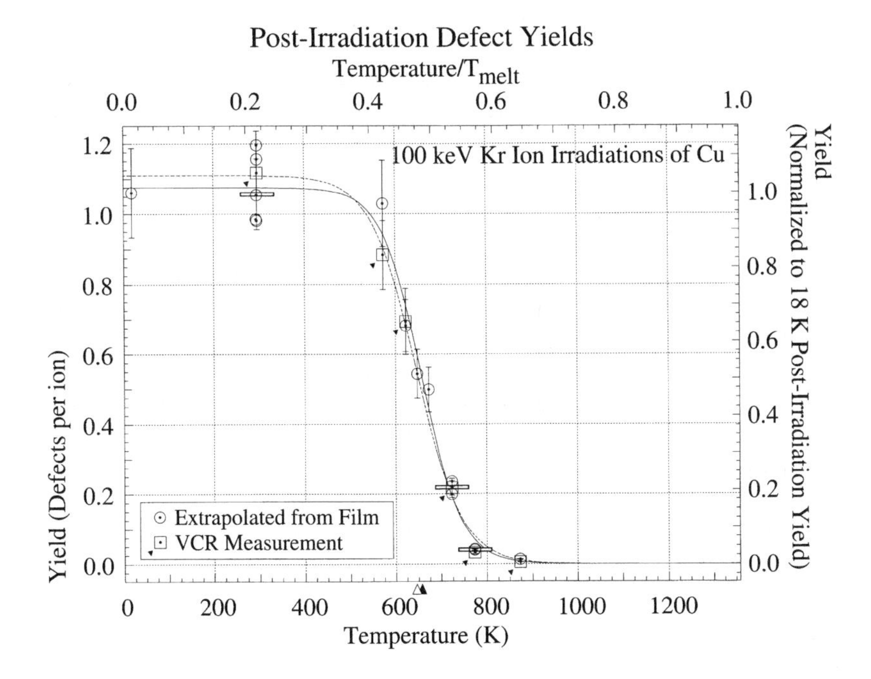

Figure 9.12. *In situ* measurements of the post-irradiation defect yield in copper after irradiation with 100 keV Kr^+ ions, as a function of the irradiation temperature. The yields are measured in two independent ways: □ direct measurements from video-cassette-recorded images; O measurements from images recorded on film, with the effect of isothermal anneal losses eliminated. This experiment shows that cascade collapse in copper is temperature dependent. From Daulton *et al* (2000).

dynamic aspects of these processes. Events could be recorded on video with a time resolution of 33 ms. Figure 9.13 shows a sequence of unprocessed video frames during 200 keV Xe^+ irradiation at room temperature of a 10–20 nm thick gold specimen. These are bright-field images recorded at an underfocus of about 1 μm. Frame 1 shows a hole (A) at the top left of the field of view, produced by an earlier ion impact. Between frames 1 and 3, a single-ion impact has occurred at B, creating a new hole and partially filling hole A with ejected material. Between frames 79 and 80, hole B has been further enlarged and hole A has become even smaller.

In situ microscopy has also been used to study the creation and annihilation of craters in gold and other materials (Donnelley and Birtcher 1997, Birtcher and Donnelley 1998). Figure 9.14 shows examples of surface craters produced in a thin gold foil by the impact of single Xe^+ ions. Craters are produced by approximately 2–5% of impinging Xe^+ ions in the energy range 50–400 keV,

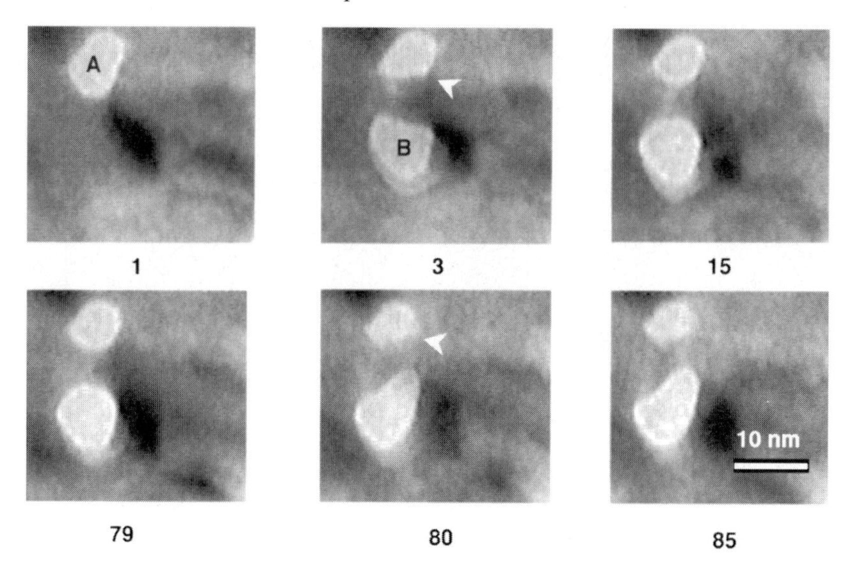

Figure 9.13. A dynamic experiment showing hole formation in a thin gold film under 200 keV Xe^{+} irradiation at 300 K. The changes occurring between frames 1 and 3, and 79 and 80, are probably produced by single ion impacts. Bright-field image with a defocus of about 1 μm. From Birtcher and Donnelley (1996).

and can be as large as 12 nm at the highest energies. Such craters have been predicted by molecular dynamics simulations of Averback and Ghaly (1994). They occur when the quasi-liquid region produced at the core of a cascade in the thermal-spike phase intersects the foil surface. This results in the plastic flow of material onto the surface in a volcano-like manner. Larger craters frequently have expelled matter associated with them, often taking the form of mounds (figure 9.14(*c*)) or lids (figure 9.14(*d*) and (*e*)). Sometimes isolated particles are seen, which may result from the ejection of a solid plug, which comes to rest some distance from the crater (figure 9.14(*f*)). At elevated temperatures craters persist only for short times. Two annihilation mechanisms were identified. Some craters annihilate in discrete steps due to plastic flow induced by subsequent ion impacts, including impacts which do not themselves produce craters. Craters also anneal in a continuous fashion due to surface diffusion. In gold the surface diffusion process responsible for thermal annealing of craters was found to have an activation energy of 0.76 eV.

Example 4: Void swelling under ion irradiation

In example 3 of the previous section, void-swelling studies under electron irradiation were described. It was pointed out that an important aspect of fast-neutron irradiation, the production of high-energy displacement cascades, is not

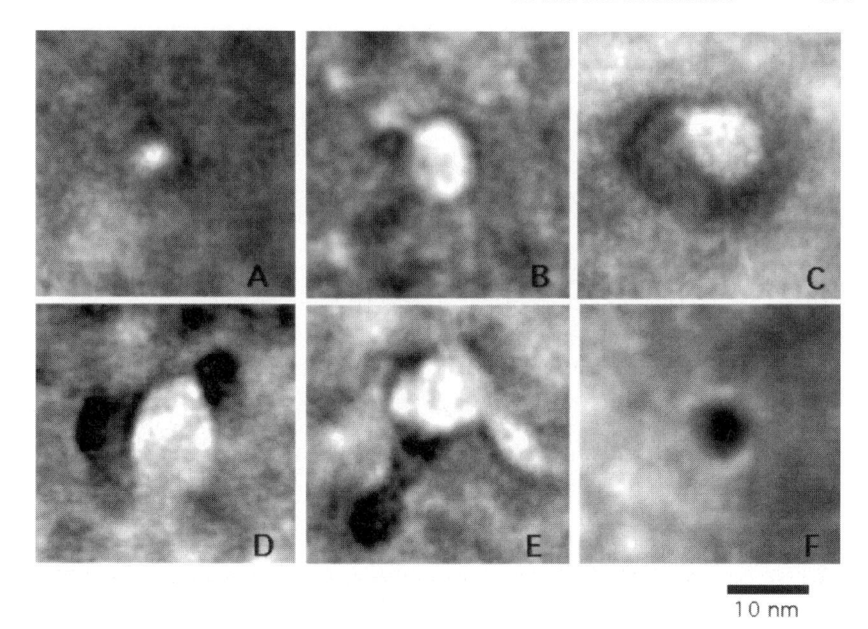

Figure 9.14. Images (digitized from video recordings) resulting from the impact of single Xe^+ ions on a gold foil: craters produced by (*a*) 50 keV, (*b*) 200 keV, (*c*), (*d*) and (*e*) 400 keV Xe^+ ions; (*f*) an isolated particle which appeared on the surface between successive video frames during iradiation with 400 keV Xe^+ ions. From Donnelley and Birtcher (1997).

well simulated by electron irradiation because the knock-on energies are too low. A better simulation of neutron damage can be made by employing heavy-ion irradiation.

An example of an *in situ* high-energy ion-irradiation experiment to simulate high-dose neutron irradiation of an advanced low-swelling stainless steel is illustrated in figure 9.15. A very high dose (200 dpa) can be achieved by 1.5 MeV Kr^+ ion irradiation at elevated temperature (650 °C) in the HVEM Tandem Facility at Argonne in a few hours, equivalent to several decades of neutron irradiation. In such an irradiation of a TEM thin sample to high dose, the effect of nearby surfaces is suppressed by the very high density of irradiation-produced sinks for mobile point-defects, thus more closely simulating a neutron irradiation of a bulk sample. The higher dose rate in the ion irradiation does produce an upwards shift in the temperature of peak swelling. However, the suppression of void swelling by the addition of titanium, seen previously in lower-dose neutron irradiation (see figure 6.1), is predicted in the results shown in figure 9.15.

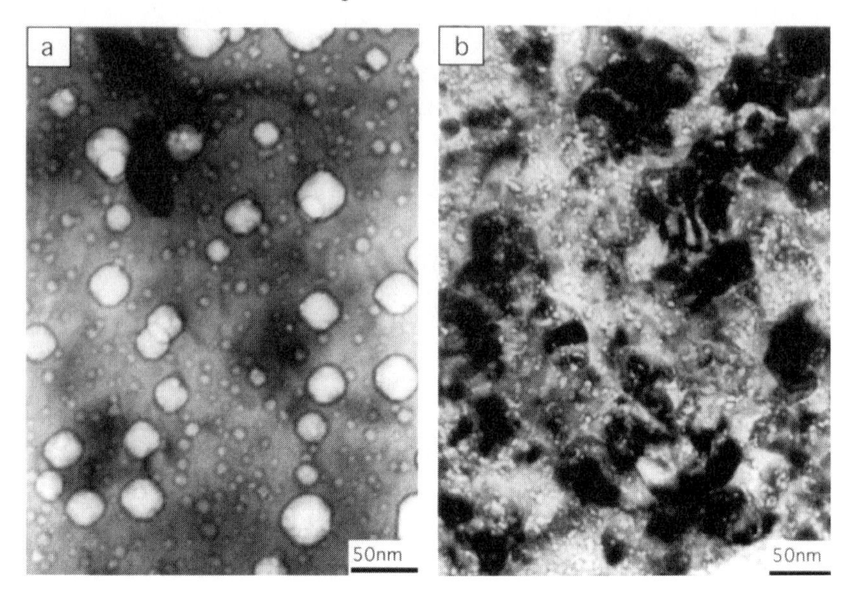

Figure 9.15. (*a*) Voids in a 16Cr–15Ni–3Mo stainless steel and (*b*) in the same alloy with 1% Ti addition. Irradiation by 1.5 MeV Kr^+ ions at 650 °C to a dose of 200 dpa. Note that the addition of Ti suppresses void swelling. Compare with figure 6.1. From work by Sagaradze *et al* (1999).

9.4 Future possibilities

In situ microscopy undoubtedly has a bright future. Ever-improving image acquisition and recording possibilities will enable a wider range of experiments to be performed. Of particular present interest to the radiation-damage community is the possibility of combining *in situ* irradiations with *in situ* straining. This will open a new avenue for investigating the important embrittlement problem in reactor pressure-vessel steels. We envisage experiments to investigate directly the interactions of dislocations with both copper-rich precipitates and other dispersed obstacles, such as small, possibly submicroscopic, dislocation loops introduced by irradiation. Such experiments will test current theories of hardening (such as the Russell–Brown modulus hardening mechanism for coherent copper-rich precipitates) and give a handle on the intractable matrix-damage problem (see section 5.2.1).

In the medium to long term, the advent of a new generation of microscopes with spherical aberration correctors is likely to lead to new possibilities for *in situ* studies. At present, high resolution (with a point resolution less than, say, 0.2 nm) can only be achieved in microscopes with an objective lens of low spherical aberration (typically, $C_s < 1$ mm). Of necessity, such lenses have small pole-piece gaps, which are not compatible with heating, cooling or straining stages or *in situ* irradiation. This restriction will disappear when spherical aberration correctors are available, leading to the possibility of combining high-resolution imaging and small-probe operating modes such as microdiffraction with environmental control. This will be exciting indeed.

Chapter 10

Applications of analytical techniques

Analytical transmission electron microscopy offers the possibility of simultaneously generating chemical and microstructural information from the same local region of a specimen. The analytical techniques seek to utilize the wide range of signals generated by the incident electron beam. The more common techniques include energy-dispersive x-ray spectrometry (EDX), electron energy-loss spectroscopy (EELS), energy-filtered imaging and convergent-beam diffraction (CBED). All of these techniques may be available in a single instrument. In addition, the imaging and diffraction capabilities of TEM are retained, so that chemical and microstructural information can be correlated.

Analytical microscopes fall into two main categories: dedicated analytical microscopes equipped with field-emission electron guns; and instruments based on standard TEMs with conventional LaB_6 or tungsten hairpin filaments. The analytical techniques all involve the production of a fine electron probe. Field-emission guns are capable of putting a large current into a very small probe (e.g. 0.5 nA into a probe of diameter 0.2 nm is possible with the latest instruments), and so dedicated analytical instruments are designed and configured so that such small probe sizes can be achieved and usefully employed. This involves the use of low-aberration lenses and fine control of probe position. With conventional electron sources, it is difficult to produce probes of diameter <10 nm with useful current, and so the spatial resolution of dedicated analytical instruments in analytical modes is generally better. Field-emission guns require high or ultra-high vacuum conditions in the gun chamber, and so dedicated analytical microscopes are less routine in operation and more expensive than conventional instruments. The fine electron probes in both conventional and the latest dedicated instruments (so-called FEGTEMs) are achieved by the use of condenser-objective lenses. Such instruments can be operated in imaging mode in the same way as a standard TEM, and all of the techniques of diffraction contrast and high-resolution imaging described earlier in this book can be applied.

The first generation of dedicated analytical microscopes were scanning transmission electron microscopes (FEGSTEMs) such as the Vacuum Generators'

HB 501. Conventional analytical electron microscopes are also often equipped with scanning coils and so can also be used in STEM imaging mode. STEM combines features of both conventional transmission and scanning electron microscopy. As in scanning electron microscopy (SEM), the image is built up sequentially by scanning a fine electron probe over the specimen in a raster. However, the specimen is a thin foil as in the conventional TEM (CTEM), and the *transmitted* electrons provide the signal used to generate the image. In fact, it can be shown that STEM images obtained using a conventional on-axis detector show essentially similar contrast to CTEM images. Dynamical contrast features such as bend contours and thickness fringes may appear blurred because of the large collection angle of the transmitted-electron detector (effectively resulting in a range of excitation errors s_g). However, it is, for example, possible to image dislocations, and to analyse their Burgers vectors using the standard $\boldsymbol{g} \cdot \boldsymbol{b} = 0$ invisibility criterion discussed earlier in this book. Such images do not provide *extra* information over CTEM images, and so confer no particular advantage[1]. STEM imaging is nearly always combined with microanalytical techniques such as EDX or EELS in order to select the specimen point of interest for analysis. The probe can be stopped on this area and point analysis performed. Moreover, it is the *only* imaging mode in dedicated STEMs equipped with field-emission guns.

The great power of analytical electron microscopy has resulted in its widespread use in materials science, and several excellent books are available which describe its principles and techniques in detail (e.g. *Practical Analytical Electron Microscopy in Materials Science* by D B Williams (1983)). For this reason we shall limit this chapter to a description of a few key experiments in radiation damage. These experiments demonstrate the capability of analytical microscopy to give information, sometimes quantitative, on radiation enhanced or induced precipitation, phase separation and segregation phenomena. Further examples, particularly of the work of the Oak Ridge group, may be found in a recent review of the application of analytical electron microscopy to radiation damage by Kenik (1994).

10.1 Examples of the use of analytical techniques in radiation damage

Example 1: Identification of copper-rich precipitates in neutron-irradiated ferritic alloys

High-resolution studies of the structure of copper-rich precipitates in irradiated model alloys have been described earlier in this book (section 9.3). It can be confirmed that such precipitates really are copper-rich by analytical electron microscopy. Particularly convincing perhaps are energy-loss images such as that

[1] An exception is high-angle annular dark-field imaging, which is discussed in section 10.4, example 4 below.

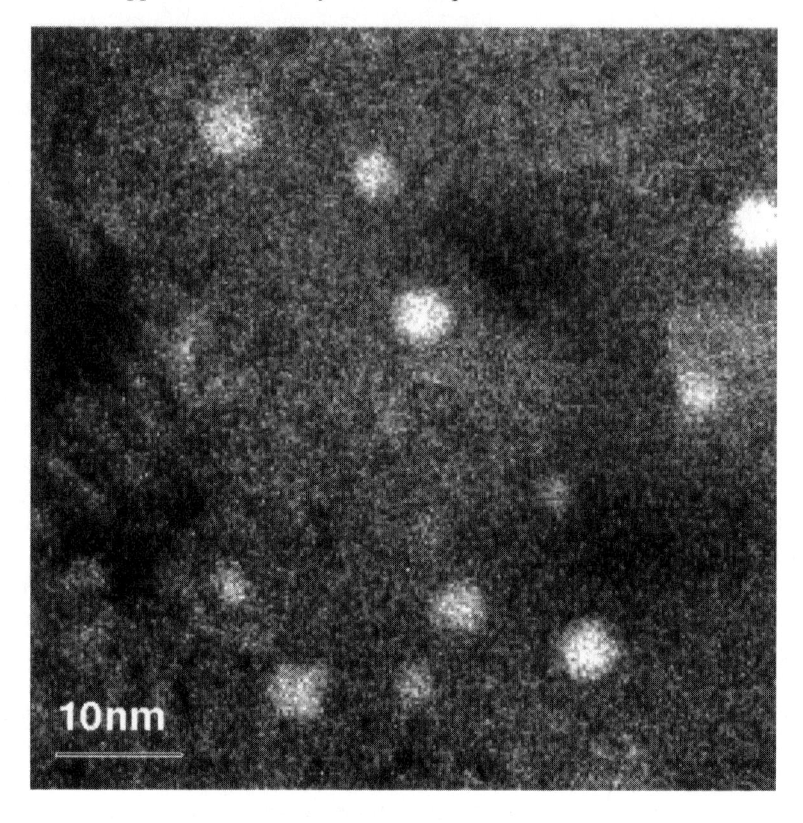

Figure 10.1. 'Jump ratio' image showing small copper-rich precipitates in an Fe–1.3wt%Cu model alloy irradiated with neutrons to a dose of 8.61×10^{-3} dpa at a temperature of ~270 °C. See text for details. Unpublished work of A Nicol, M L Jenkins and R Dunin-Borkowski.

shown in figure 10.1. This micrograph was obtained in a dedicated FEGTEM equipped with a Gatan imaging filter[2]. It shows small copper-rich precipitates in an Fe–1.3wt%Cu model alloy irradiated with neutrons to a dose of 8.61×10^{-3} dpa at a temperature of ~270 °C. This image was formed essentially only by electrons which had lost a specific amount of energy, in this case corresponding to a small energy window centred at $\Delta E = 1000$ eV, just above the copper L ionization edges at 931 and 951 eV. The background has been reduced by dividing the raw energy-loss image by a similar image obtained using an energy window well below the Cu_L edges, at $\Delta E = 915$ eV. In such a 'Cu jump ratio' image, only copper-rich regions shine out brightly. The precipitates, of size 2–6 nm, are clearly visible.

[2] The probable advantage of energy-filtering in weak-beam images, using *zero-loss* electrons, was mentioned in section 3.3. See figure 3.22.

Studies of this sort have only started to be undertaken recently and, as far as we are aware, energy-filtering has not been applied to real irradiated pressure-vessel steels. This is likely to change in future.

Example 2: Identification of radiation-induced and radiation-enhanced phases in a neutron-irradiated austenitic steel

Precipitation under irradiation in austenitic steels often occurs copiously and at much lower temperatures than in thermally-aged material; and this has important consequences if the material is used in an irradiation environment. The identification of such phases and the determination of their compositions by analytical electron microscopy are difficult and time-consuming but can be achieved even with a non-dedicated analytical electron microscope. Essentially the idea is to combine EDX, which gives semi-quantitative chemical information on heavier elements, with convergent-beam and microdiffraction, which give structural information and lattice parameters. This enables the various precipitate types, which co-exist in a given material, to be identified. The EDX spectra for precipitates of each given type show characteristic features, and once these features have been established they can be used as a 'fingerprint' to rapidly identify other precipitates of the same type. In this way the full precipitate distributions can be established without the composition of each precipitate being determined explicitly. The ease of identification of given particles depends on the foil thickness and the particle size, but $\sim$50 nm particles can readily be identified, and identification of considerably smaller particles is possible.

A catalogue of convergent-beam diffraction patterns and EDX spectra has been published by the Bristol Microscopy Group (Steeds and Mansfield 1984) and this allows the quick and routine identification of many of the phases found in complex multiphase alloys.

Williams *et al* (1982) describe the use of the procedure for identifying phases in a neutron-irradiated niobium-based austenitic stainless steel, FV 548. The most commonly occurring precipitates which have been identified in irradiated FV 548 alloy, together with their characteristic EDX fingerprints (see figure 10.2) will now be listed. Also shown in figure 10.2 for comparison are (*a*) a 'hole'-count spectrum, showing Mn_K peaks arising from the decay of the Fe^{55} isotope produced by neutron irradiation and (*b*) a matrix spectrum, obtained from a region not containing precipitates.

(i) Niobium carbide, NbC. This is the principal carbide in FV 548. The distribution of these carbides is significantly modified by irradiation, with a considerable increase in the number of fine intergranular precipitates. This phase is believed to be responsible for suppressing void swelling in FV 548 by acting as an efficient site for recombination of point-defects. The x-ray spectrum (figure 10.2(*c*)) consists of strong Nb_K and Nb_L peaks, and a very low Si_K peak superimposed on the matrix.

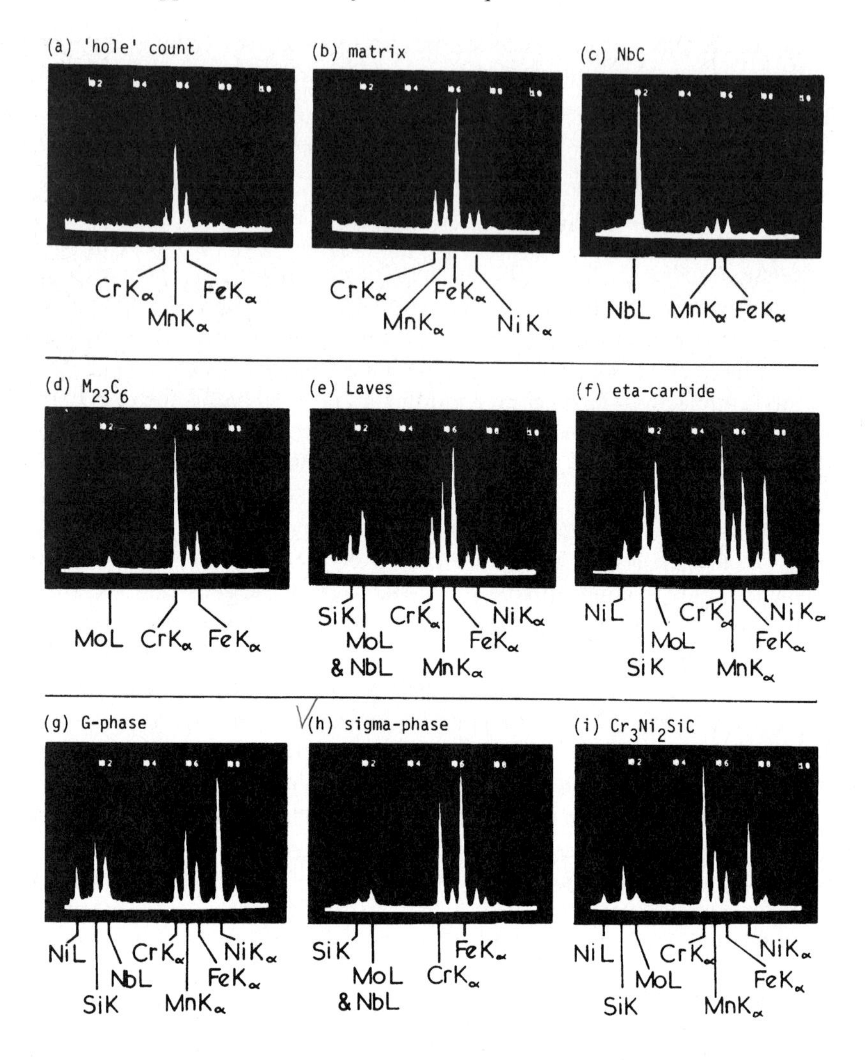

Figure 10.2. Typical EDX spectra used to 'fingerprint' precipitates in austenitic stainless steels: (*a*) hole-count, showing a strong Mn_K peak arising from residual activity of the whole specimen; (*b*) matrix; (*c*) niobium carbide; (*d*) carbides of type $M_{23}C_6$; (*e*) Laves phase; (*f*) η-carbide (M_6C); (*g*) G-phases, which are silicides of type $T_6Ni_{16}Si_7$; (*h*) sigma phase; (*i*) Cr_3Ni_2SiC. From Titchmarsh and Williams (1981).

(ii) Carbides of type $M_{23}C_6$ with the fcc structure. $M_{23}C_6$ is found in both unirradiated and irradiated FV 548, with similar distributions, and is compositionally very stable. The x-ray spectrum of $M_{23}C_6$ carbide is distinguished by an increased Cr_K peak and the absence of silicon, which prevents confusion with the Laves phase or η-carbide (M_6C), see figure 10.2(*d*).

(iii) Laves phase, which is a hexagonal phase with a composition based on Fe_2Mo or Fe_2Nb. This phase is only seen in FV 548 in specimens irradiated above the void-swelling temperature range. It has a similar Ni content and slightly increased Cr content compared with the matrix, and generates noticeable Si_K and $(Nb_L + Mo_L)$ peaks. The $Si_K/(Nb_L + Mo_L)$ ratio is typically ≤ 0.5 (figure 10.2(*e*)).

(iv) η-carbide (M_6C). This occurs only in irradiated FV 548. The phase has the diamond-cubic structure. η-carbide has greatly increased silicon, (niobium + molybdenum), and nickel concentrations compared with the matrix. It is distinguished from the Laves phase by having a $Si_K/(Nb_L + Mo_L)$ ratio of about 0.75 to 1, and a larger Ni:Fe ratio (figure 10.2(*f*)). The identification of M_6C is discussed explicitly later.

(v) G-phases, which are silicides of type $T_6Ni_{16}Si_7$, where T is a transition element. The main G-phase found in irradiated FV 548 containing niobium is $Nb_6Ni_{16}Si_7$,which has an fcc structure. It was found in significant quantities only in material irradiated in the aged condition. G-phase has an increased nickel content compared with the matrix, and a $Si_K/(Nb_L + Mo_L)$ ratio ≥ 1 (figure 10.2(*g*)).

(vi) Sigma phase, containing iron, chromium, molybdenum and silicon, with very low levels of Ni, which was found as coarse grain-boundary particles in specimens irradiated at high temperatures (738 °C, well above the temperature range for void swelling). This has a significantly higher chromium content than other phases (figure 10.2(*h*)).

(vii) Cr_3Ni_2SiC. This is a second form of M_6C, chemically different from the η-carbide described earlier. It shows larger Cr, Ni and Si peaks than the pure matrix, but little Mo or Nb. It is distinguished from the sigma phase and $M_{23}C_6$ by the Ni/Fe ratio being greater than in the matrix.

In order for this 'fingerprinting' technique to be used reliably, it is first necessary to identify unambiguously precipitates of the different types. This requires correlation of chemical and structural information. The identification of the radiation-induced, silicon-rich, η-carbide phase in neutron-irradiated FV 548 is discussed here as an example. The composition of the phase may be estimated from the EDX spectrum of a particle in a thin foil shown in figure 10.2(*f*) using standard techniques of quantitative analysis. The proportions of the heavier elements are found to be (in atomic %): 21 Ni, 21 Si, 28 (Mo + Nb), 18 Cr, 12 Fe. The Mo:Nb ratio varies from ~2:1 to ~1:2. Note that no information is available from EDX about the possible carbon content of the precipitate, and none would be expected. Light element analysis using EDX is difficult or impossible, because of the low fluorescent yield and the high absorption of the soft characteristic x-rays in the specimen and inactive regions of the detector. In the present experiment a beryllium thin-window x-ray detector was used, and carbon x-rays would have

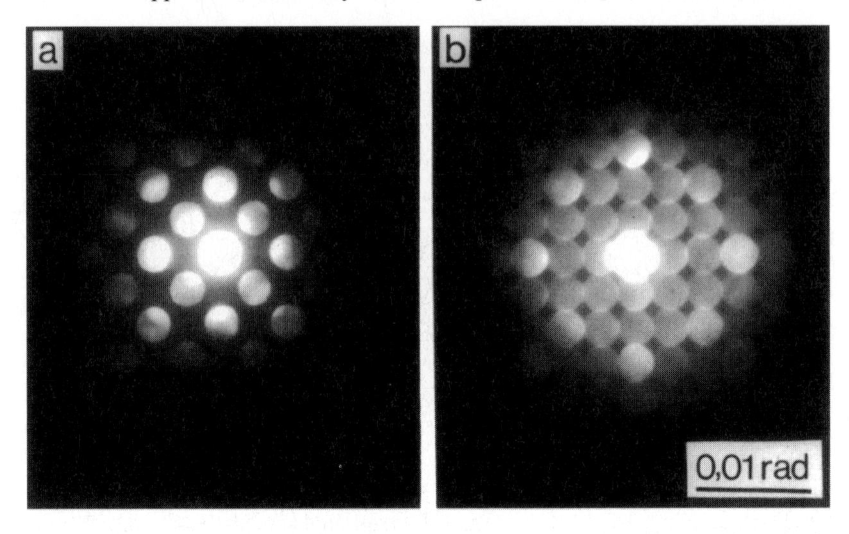

Figure 10.3. ⟨100⟩ zone axis microdiffraction patterns for (*a*) η-carbide (M_6C) with the diamond-cubic structure, and (*b*) $M_{23}C_6$ with the fcc structure. From Williams and Titchmarsh (1979).

been almost completely absorbed before reaching the active region of the detector. However, even with a windowless detector it is unlikely that quantitative analysis would have been possible, since carbon contamination is usually present as a surface film on thin-foil specimens. The presence or absence of carbon (or other light elements) in precipitates must therefore be deduced indirectly unless techniques sensitive to light elements (such as EELS) are also used.

Figure 10.3 shows a ⟨100⟩ zone-axis microdiffraction pattern from a particle of the phase compared with a similar pattern from an fcc $M_{23}C_6$ precipitate. This diffraction pattern, together with other micro- and convergent-beam diffraction patterns taken at different orientations (see Steeds and Mansfield 1984), show that the new phase is cubic with lattice parameter $a = 1.09 \pm 0.1$ nm. This value is close to those of $M_{23}C_6$ ($a = 1.125$ nm) and the G-phase (1.064 nm). However, the absence of {200} reflections indicates that the structure is diamond cubic rather than fcc, so these phases can be excluded. The data are consistent with the new phase being a carbide of type M_6C, which does have the diamond-cubic structure and is known to exist with widely differing compositions and a small variation in lattice parameter depending on composition. Silicon is known to enter the M_6C phase either interstitially or substitutionally. This phase is not found thermally in FV 548, and its presence in irradiated FV 548 must therefore be a consequence of the irradiation. It probably results from the non-equilibrium segregation of solutes such as silicon and nickel to sinks under irradiation, see example 3 below.

In summary, the main effects of irradiation in FV 548 are (i) to enhance the

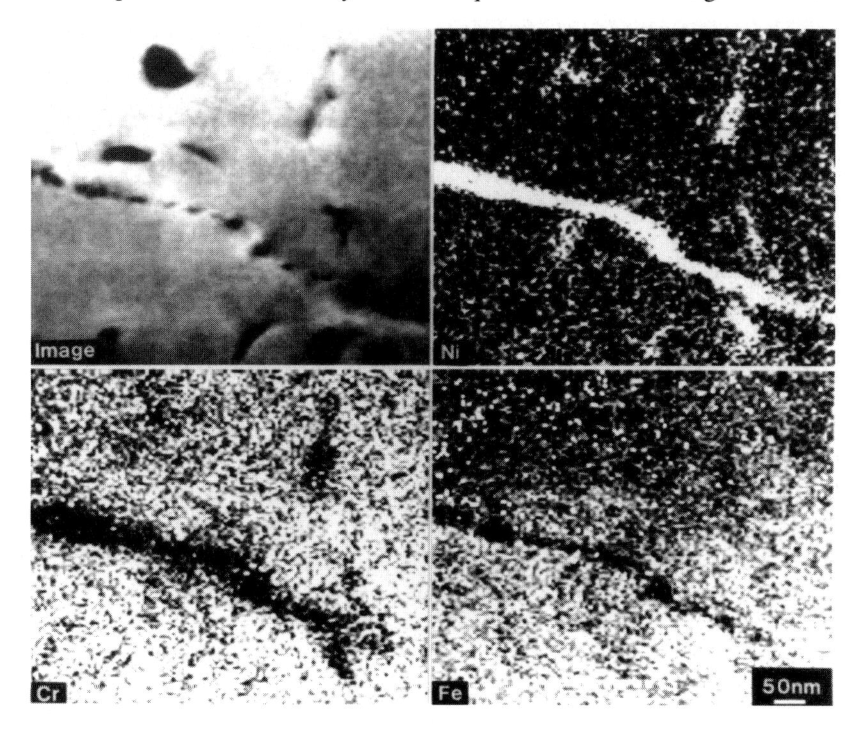

Figure 10.4. FEGSTEM investigation of segregation at a dislocation in an FV 548 alloy irradiated to a dose of 5 dpa with 46.5 MeV Ni^{6+} ions at 600 °C showing a bright-field STEM image, and x-ray maps for nickel, chromium and iron. The dislocation is seen to be enhanced in nickel, and depleted in chromium and iron. From Shepherd (1990).

formation of the η-carbide and, under some conditions, the G-phase, sigma phase and Laves phase; and (ii) to produce fine intergranular distributions of NbC which play a key role in suppressing void swelling.

Example 3: Radiation-induced segregation in irradiated stainless steels

Irradiation sets up point-defect fluxes to sinks such as grain boundaries. The different atomic species present in an alloy respond differently to these fluxes—some species tend to move with the point-defect fluxes towards sinks, others move away. The resulting radiation-induced segregation can cause significant variations in composition near sinks, and may therefore have severe deleterious consequences for the mechanical properties of the material. A particular example is chromium depletion at grain boundaries in austenitic stainless steels, which lowers the corrosion resistance of the alloy and can lead to stress corrosion cracking.

The phenomenon may be investigated using a dedicated field-emission gun analytical microscope. The effect can be seen qualitatively by x-ray mapping, in favourable cases even for sinks such as dislocations. Figure 10.4 shows x-ray maps for nickel, chromium and iron obtained using a FEGSTEM for an FV 548 alloy irradiated to a dose of 5 dpa with 46.5 MeV Ni^{6+} ions at 600 °C (Shepherd 1990). The dislocation at the centre of the field of view, imaged in the bright-field STEM image of figure 10.4(*a*), is seen to be strongly enhanced in nickel (figure 10.4(*b*)), and depleted in chromium and iron (figure 10.4(*c*) and (*d*)). These changes in composition appear to have occurred within about 20–30 nm of the dislocation. In this case, the segregation is so strong that it is easily detectable for a dislocation lying in the plane of the foil, even though this is not be the most favourable geometry (viewing the dislocation end-on would give a bigger signal).

Most attention has been paid to radiation-induced segregation to grain boundaries, both because of its technological importance and the possibility of quantitative analysis. The boundary can be oriented edge-on and a series of EDX spectra acquired across the boundary. With a dedicated FEGSTEM the probe size can be less than 1 nm, and the positioning of, and spacing between, analysis points can be controlled to a similar level of accuracy. Examples of experimental measurements of the chromium concentration across two boundaries in a neutron-irradiated stainless steel are shown in figure 10.5 taken from Titchmarsh and Dumbill (1996). These measurements show an interesting and unexpected feature: although chromium is depleted from a zone of width 20–30 nm adjacent to the grain boundary, there appears to be a peak in chromium concentration at the boundary itself. Such a maximum is not seen when similar experiments are performed with a larger probe size using a conventional analytical microscope, suggesting that boundary segregation may sometimes be concealed by inadequate electron probe resolution.

Such observations have led Titchmarsh and Dumbill (1996) to examine by Monte Carlo modelling the various factors (such as the electron probe width, the foil thickness and the segregation profile) which influence the shape of the experimental profiles such as those shown in figure 10.5. They were able to demonstrate good agreement between the experimental profile of figure 10.5 and the Monte Carlo simulation on the assumption that 0.8 monolayers of chromium were segregated to the boundary, together with assumed values of the other parameters shown in figure 10.6. Titchmarsh and Dumbill emphasize that the figure of 0.8 monolayers of chromium probably does not give a unique match—it may be possible to get an acceptable match by some other combination of parameters. In order to improve the estimate, accurate measurements of foil thickness and probe size would be necessary. The simulations show, however, that the apparent segregation of chromium to the boundary is real. It probably occurred prior to irradiation during solution treatment of the alloy by non-equilibrium segregation.

More recently, further progress in the quantitative analysis of STEM-EDX compositional profiles in neutron-irradiated stainless steels has been made, by applying the technique of multivariate statistical analysis (Titchmarsh and

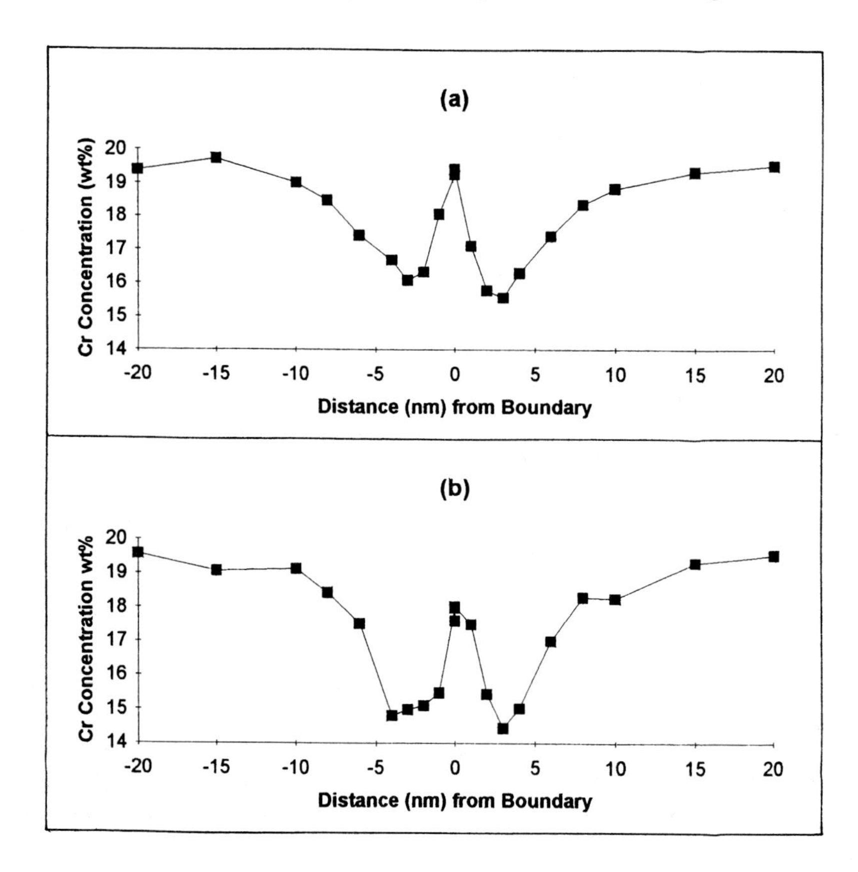

Figure 10.5. Experimental FEGSTEM measurements of Cr concentration at two boundaries in neutron-irradiated stainless steels. Each profile clearly demonstrates the presence of a depletion profile adjacent to the boundary, in addition to segregation at the boundary. From Titchmarsh and Dumbill (1996).

Dumbill 1997). This sophisticated technique allows the maximum amount of information to be extracted from the experimental spectra. It is particularly useful for revealing small peaks close to the sensitivity limit of EDX. It was found possible by this technique to deconvolute the atomic concentration profiles in the boundary region. Experimental data for Cr, Ni and Si, measured across a grain-boundary in a neutron-irradiated model stainless steel, are shown in figure 10.7 as full squares. The concentration profiles derived from these experimental data, modelled after multivariate statistical analysis, are shown as full curves. In order to check the consistency of the analysis, these calculated elemental profiles were used as input data for Monte Carlo calculations, for the same conditions as the actual experiment. The results of the Monte Carlo calculations are shown as

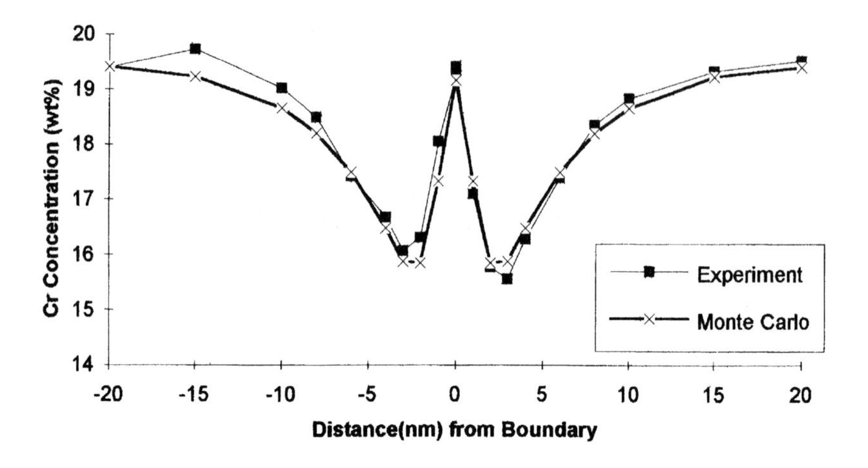

Figure 10.6. Comparison of Monte Carlo results fitted to the experimental data of figure 10.5. Parameters used: foil thickness $t = 50$ nm; probe diameter $= 4.0$ nm; width of profile $L = 4.5$ nm; initial concentration at interface $C_s = 12\%$; bulk matrix concentration $(C_s + C_m)$ with $C_m = 7.5\%$; and 0.8 monolayers Cr segregation. From Titchmarsh and Dumbill (1996).

crosses in figure 10.7, and clearly show good agreement with the experimental data. The analysis confirms the segregation of chromium to the boundary plane, and suggests that both nickel and silicon are considerably depleted on the boundary plane, contrary to previous expectation. Further details of this work are beyond the scope of the present book.

Example 4: Damage nucleation at grain boundaries in silicon investigated using high-angle annular dark-field imaging

High-angle annular dark-field or 'Z-contrast' imaging has been largely developed by S J Pennycook (Pennycook and Jesson 1991, Pennycook 1992). The technique was originally developed for dedicated FEGSTEM instruments, but can also be used in FEGTEM instruments equipped with a scanning attachment. It involves the use of an annular solid-state detector, which collects high-angle, elastically scattered electrons, in an STEM imaging mode. The signal produced by such electrons can be shown to be proportional to the average atomic number of the region sampled by the probe, yielding local compositional information. The very fine probes in a FEGSTEM allow such Z-contrast images to be achieved at atomic resolution. Unlike conventional high-resolution structural images, Z-contrast images do not rely on phase contrast and are relatively insensitive to focus. Their interpretation is therefore relatively straightforward. An equivalent technique in FEGTEM instruments utilizes hollow-cone illumination, achieved by the use of an annular condenser aperture (Bentley *et al* 1990). This does not

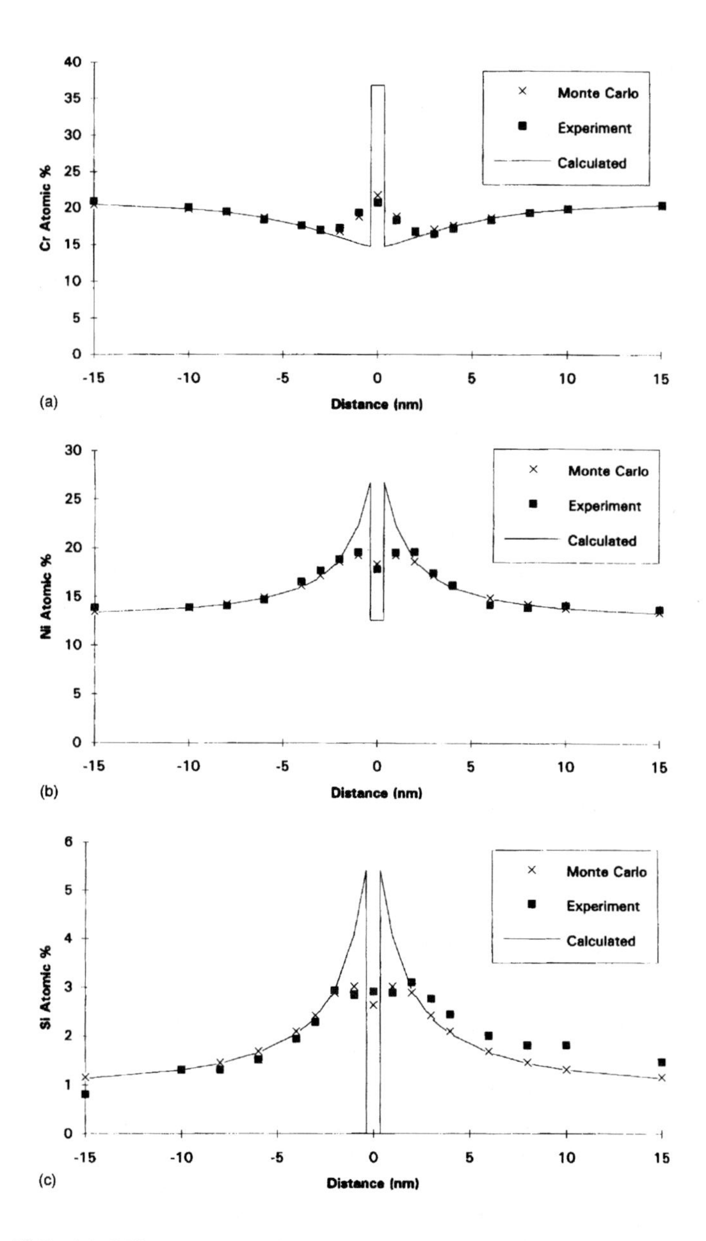

Figure 10.7. Modelling concentration profiles of (*a*) Cr, (*b*) Ni, and (*c*) Si at a grain boundary in an irradiated stainless steel: ■, experimental measurements using FEGSTEM; full line, concentration profiles derived from these experimental spectra, modelled after multivariate statistical analysis; $\times$, Monte Carlo simulations of the experimental data using these concentration profiles. From Titchmarsh and Dumbill (1997).

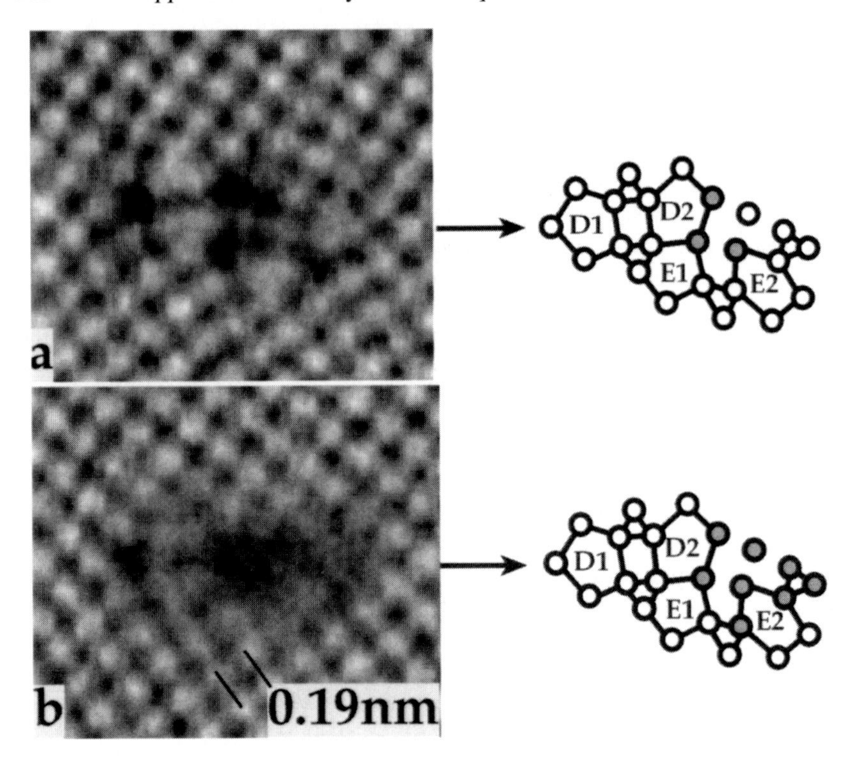

Figure 10.8. Z-contrast images and derived structures of the $\Sigma = 25\{710\}\langle 001\rangle$ symmetric tilt boundary at two stages of exposure to 300 keV electron irradiation: (*a*) a nearly unaffected core with all the columns visible but those shaded in the schematic diagram showing reduced intensity; (*b*) a partially affected core with several columns appearing darker. From Maiti *et al* (1999).

require beam scanning.

The Z-contrast technique suffers from a lack of sensitivity to defect scattering in the presence of strong crystalline scattering. This makes it largely unsuitable for characterization of small (<10 nm) defect clusters of sizes considerably less than the foil thicknesses used for structural resolution (10–100 nm). For this reason, the technique has not been employed widely for radiation-damage studies.

The technique can, however, be used to image defect structures with planar geometry extending from the top to the bottom of the foil, such as grain boundaries. A recent application in radiation damage is the study of damage nucleation and vacancy-induced structural transformation in silicon grain boundaries by Maiti *et al* (1999). The structure of symmetric $\Sigma = 25$ tilt boundaries in silicon in a projection normal to the $\langle 001\rangle$ tilt axis consists of periodically repeated units consisting of two extended cores, with one being the mirror image of the other, and separated by regions of perfect crystal. Figure 10.8 shows Z-contrast images of one such extended core, imaged using a dedicated FEGSTEM operating at 300 keV. Each atomic column is visible as a bright dot. At an early stage of exposure to the electron beam, figure 10.8(*a*), all columns are visible, and are consistent with the model of the boundary structure indicated in the figure. The shaded columns are less bright and so show early evidence of damage nucleation. Under continued exposure to the electron beam, damage accumulates preferentially on these and neighbouring atomic columns, and is visible in figure 10.8(*b*) as columns of considerably reduced brightness. Computer simulations showed that the columns where damage first nucleates correspond to sites where the formation energies of vacancies and vacancy complexes are very low.

10.2 Future trends

Several developments are likely to make analytical electron microscopy increasingly important in radiation-damage studies in the next few years. Already prototype spherical-aberration correctors have been designed, and these will allow considerably smaller electron probes than are currently possible (Krivanek *et al* 1999). Sub-ångström probes seem feasible. Smaller probes will allow, for example, more accurate and quantitative measurements of grain-boundary segregation as well as improved resolution in high-angle annular dark-field imaging. A new generation of microcalorimeter-based x-ray detectors looks set to revolutionize energy-dispersive x-ray spectroscopy (Newbury *et al* 1999). Such detectors offer much reduced peak widths, eliminating peak-overlap problems, as well as an increase of over an order of magnitude in peak-to-background ratios. Combined, these developments will probably lead to near-atomic resolution and sensitivity, and will open new opportunities, for example the measurement of chemical peak shifts by EDX, giving information on local atomic environments. EELS and energy-selected imaging are also likely to become of increasing importance. Already electronic bonding studies by EELS are almost routine, and the problems of interpretation and modelling are being addressed vigorously. Further advances of these techniques may be aided by the development of energy monochromators to reduce the electron energy spread at the gun to perhaps 0.1–0.2 eV (Mooh and Kruit 1999). Also important will be the likely development of more stable power supplies (from one part in 10^6 to a few parts in 10^7) and improvements in spectrometer design which promise to improve the energy resolution from about 0.4 eV to 0.1 eV. Together, such advances in EDX and EELS instrumentation and theory are likely to allow measurements of the structure, chemistry and cohesive strength of real structural materials on a near-atomic scale. Clearly this will have a major impact in many areas of irradiation damage, especially perhaps in structural integrity studies.

Chapter 11

Radiation damage in amorphous glasses

11.1 Property changes caused by irradiation

The leading proposal for the containment of high-level radioactive waste for geological time spans is to incorporate it into a compact solid form. A class of material which offers one of the best chances to accomplish this difficult feat is glass. In assessing the suitability of candidate glasses for this application, the main criterion is the stability of the structure and the chemistry of the glass under internal irradiation exposure from decay of the radioactive isotopes. In this final, short chapter we summarize briefly the effects of radiation on glasses and describe briefly a new and promising technique by which radiation damage in amorphous materials can be studied in the transmission electron microscope.

The more important property changes in glasses produced by irradiation include the following ones.

(i) *Irradiation-induced volume changes.* The compaction of silica glass under neutron irradiation was first reported by Primak *et al* (1953). Subsequent studies have reported similar effects under virtually all forms of irradiation. Ion-irradiation studies have shown that the densification of silica glass is caused primarily by elastic collisional processes rather than by ionization losses (EerNisse 1974). More complex glasses may show either expansion and compaction, depending on irradiation dose and glass composition. Some of these volume or density changes have been shown to scale with ionization losses rather than collisional damage (Arnold 1986).
Various explanations have been put forward to account for the effect. These include interstitial formation (Arnold 1986), changes in the number of ring members (Devine 1994) or a simple relaxation to an unspecified more stable structure (Stoneham 1994).
The irradiation-induced volume changes are generally not large for the low doses studied, usually less than 2–3%, but in complex glasses with some inhomogeneities, volume changes may be sufficiently large to lead to

near-surface cracking at higher doses, producing the opportunity for higher leaching rates.

(ii) *Second-phase formation.* The most common observation is the formation of bubbles, most probably of helium or oxygen, which can lead to large volume changes. Bubble formation under irradiation has been observed in alkali silicate, alkali borosilicate and simulated waste glasses. Again, most forms of irradiation—gamma, electron and ion—have been shown to produce bubbles (DeNatale and Howitt 1985).
Controversy exists on the possibility that the TEM examination itself contributes to the formation of bubbles. Studies of the effect of dose and dose rate on bubble formation have been performed using different forms of irradiation to vary the dose rate (DeNatale and Howitt 1987). Temperature also plays a strong role and sample geometry may also contribute.

(iii) *Phase separation.* Radiation-induced phase separation is an important potential glass degradation mechanism in complex sodium borosilicate glasses. This separation can result in silica-poor and silica-rich phases. The first of these phases tends to support the higher concentration of fission products but unfortunately is chemically less durable. Phase separation by radiolysis under TEM irradiation has been observed in a candidate waste glass, which was probably caused or accelerated by radiation enhanced processes (DeNatale and Howitt 1982).

(iv) *Radiation-induced diffusion.* Depletion of alkali elements under irradiation-enhanced diffusion has been shown to produce regions with different damage characteristics, such as an increased tendency for oxygen bubble formation under electron irradiation.

11.2 Directions for future work

The observations summarized here have been rather limited but suggest a number of directions for future work and the following ones in particular.

- There is clearly a need for more systematic studies of irradiation effects in reference glasses as well as simulated waste glasses over a wide range of conditions. In general there is a lack of understanding of the effects of different forms of irradiation in different glass structures. In simple silica glass, the data indicate that elastic collisional processes are most important in producing damage effect (EerNisse 1974), while in alkali borosilicates and other more complex glasses, it would appear that ionization processes dominate (Arnold 1985).
- It is difficult to draw conclusions on the relative importance of collisional and ionization processes from neutron-irradiation studies, due to the unavoidable accompanying gamma exposure. It will be necessary to use electron and ion irradiations to produce a better understanding of which irradiation and damage mechanisms contribute to the various measures of damage.

- The dependences on irradiation temperature, dose and dose rate of the volume change have received little attention (Marples 1988). Clearly, more detailed experimental work is needed to relate macroscopic volume changes to structural changes under various forms of irradiation.
- It is important to understand and establish the conditions under which bubble formation occurs, and to extrapolate to the temperature and irradiation conditions of application.
- There is a need to study radiation-enhanced diffusion to internal sinks or external surfaces of any atom type of interest under various forms of irradiation, at controlled temperatures, irradiation doses and dose rates. A successful dose rate study would offer the possibility of meaningful extrapolation of rates of radiation-enhanced diffusion at the temperatures and low dose rates in application. Changes in chemistry at surfaces are important for understanding changes in dissolution or leach rates. Any tendency for segregation of radioactive species or actinides to surfaces would be potentially bad.

Clearly, future progress will demand a much better understanding of the fundamental radiation-damage processes which lead to macroscopic property changes in glasses. An understanding of the relationship between changes in atomic structure and deleterious changes in macroscopic properties, such as an enhanced rate of leaching, would give a better chance of designing a glass containing high-level nuclear waste that is resistant to degradation in the environment. At present, these fundamental processes are poorly understood (Ewing *et al* 1985). Because glasses are aperiodic in structure, an understanding of the effects of irradiation on a microscopic level is considerably more difficult to achieve than in crystalline materials.

Most microstructural work to date has used x-ray and neutron diffuse scattering data from simple silicate glass to determine radial distribution functions. Distinct peaks corresponding to particular interatomic spacings have often been identified from the known interatomic distances in similar crystalline materials. Changes under irradiation include peak shifts and changes in peak intensities (Hobbs 1994). These have been interpreted, sometimes in correlation with other measurements, to indicate changes in bond angle, changes in ring topology, the formation of internal surfaces, voids or bubble, and the formation of interstitials and vacancies. Such interpretations are subject to considerable doubt.

In order to make progress in this topic, it is necessary to develop techniques which can lead to a better understanding of what happens on the atomic level in nuclear waste glasses under irradiation. In the following section we describe briefly a TEM technique which offers this possibility.

11.3 Variable coherence microscopy (speckle patterns)

Structural measurements of amorphous materials have relied until recently on diffraction. The diffracted intensity is closely related to the Fourier transform of the pair correlation function. This approach is therefore sensitive to short-range order but is insensitive to medium-range order. The absence of information on higher-order atomic correlations is a serious hindrance to a deeper understanding of amorphous materials.

Variable coherence microscopy (VCM) is a recently established tool for quantitative analysis in the TEM of structural fluctuations in disordered materials which overcomes this limitation (Treacy and Gibson 1996). The method involves dark-field imaging using hollow-cone electron beam illumination. Such images of aperiodic structures show fine-scale contrast variations, termed 'speckle'. These fluctuations can be characterized by statistical analysis. Higher-order atomic correlations are detectable from simple statistical properties such as the variance of image intensities. Through control of the hollow-cone angle of illumination (which determines the coherence parameter), it is possible to choose the characteristic atom pair spacings which are examined. By varying the resolution of the microscope the distances over which scattering correlations exist can be probed. Statistical properties of images are measured as a function of the coherency parameter κ, which is of approximate magnitude α/λ, where α is the inner semi-angle of the hollow-cone illumination and λ is the electron wavelength.

The VCM technique has recently been applied by Gibson and Treacy (1997) to a study of the effect of annealing on the medium-range order in amorphous germanium. Plots of the normalized variance V of hollow-cone dark-field TEM images from a 20 nm thick film of germanium before and after annealing are shown in the lower part of figure 11.1. Two of the speckle images from which this figure was obtained are shown above. Both are from the unannealed sample, with the left-hand image taken at the maximum of V at $\kappa = 5.8\ \text{nm}^{-1}$ and the right-hand image at the minimum in V at $\kappa = 4.0\ \text{nm}^{-1}$. The difference in intensity variance is very clear. The maximum at $\kappa = 5.8\ \text{nm}^{-1}$ is indicative of medium-range order on a scale of $\kappa^{-1} \approx 2$ nm. Note that the speckle variance at $\kappa = 5.8\ \text{nm}^{-1}$ showed a drop by a factor of three on annealing. This is indicative of a significant *decrease* in medium-range order. Diffraction evidence as well as other experiments indicated that this decrease in medium-range order is accompanied by a modest *increase* in short-range order, consistent with a continuous random network model. Qualitatively similar results to these have also been found in amorphous silicon films. Gibson and Treacy speculate that these, at first sight contradictory, results may be explained if a granular structure on a $\sim$2 nm scale is formed during low-temperature deposition of germanium and silicon. They suppose that a large density of nucleation centres with quasi-ordered structures forms. As these impinge on one another, a large interfacial energy results from their random relative orientations. On annealing, a true random

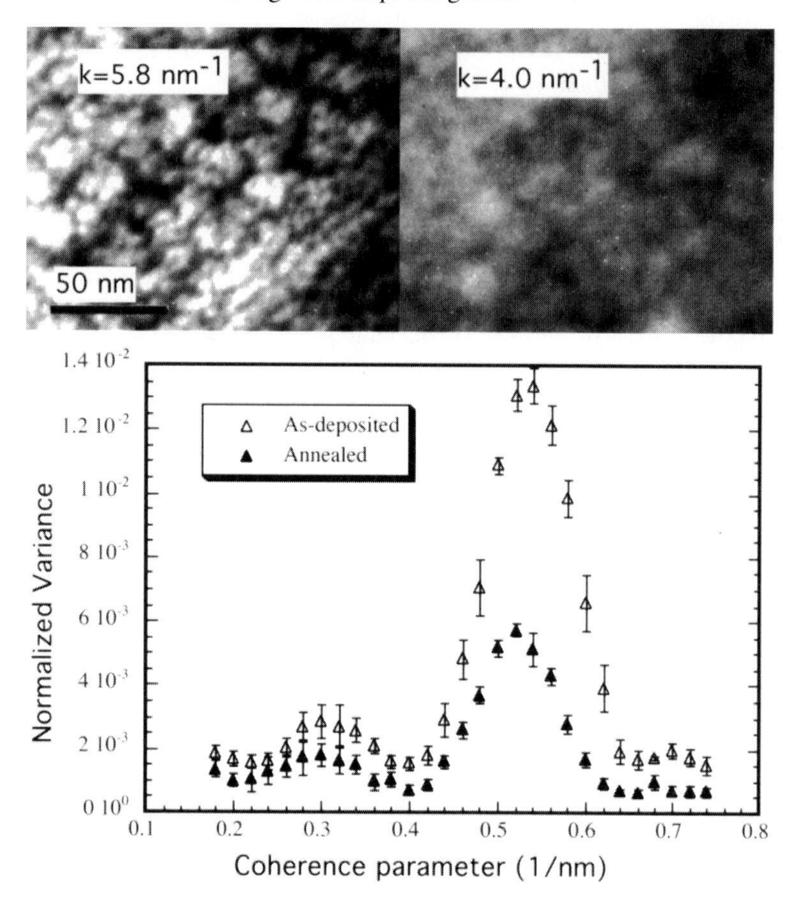

Figure 11.1. Plots of the normalized variance of hollow-cone dark-field TEM images from a 20 nm-thick amorphous germanium film before and after annealing, as a function of the coherence parameter κ. Medium-range order, exemplified by the peak at $\kappa = 5.8\ \text{nm}^{-1}$, is diminished after annealing. The variances were obtained from micrographs such as those shown above the plots. These are hollow-cone dark-field images from the same area of a 20 nm Ge film showing, on the left, maximum variance $\kappa = 5.8\ \text{nm}^{-1}$, and on the right minimum variance $\kappa = 4.0\ \text{nm}^{-1}$. From Gibson and Treacy (1997).

network forms with a uniformly distributed, lower, strain energy.

A decrease in medium-range order has also been seen in preliminary data on ion-irradiated amorphous silicon. Clearly the technique has potential for investigating structural changes produced by irradiation in other amorphous systems. For example, it would be interesting to see if similar effects occur in ion-irradiated multicomponent glasses, such that a lower energy state is produced by irradiation. If such an effect were found, it would suggest that internal irradiation from radioactive isotopes might actually result in a structure which is more stable and resistant to environmental attack.

As applied to date, the VCM technique measures all scattered electrons, some of which are degraded in energy by inelastic processes, and these constitute a major portion of the undesirable noise in the speckle patterns. This can be avoided by applying energy-filtering. Energy-filtering will have the effect of removing the degraded part of the electron energy spectrum in the image, and so improve the sensitivity of the analysis to the true average atom correlations in the structure.

Appendix A

The Thompson tetrahedron

The Thompson tetrahedron (Thompson 1953) is a useful way of visualizing close-packed planes and directions in fcc and diamond-cubic materials. The four faces of the tetrahedron correspond to the four close-packed {111} planes, and the six

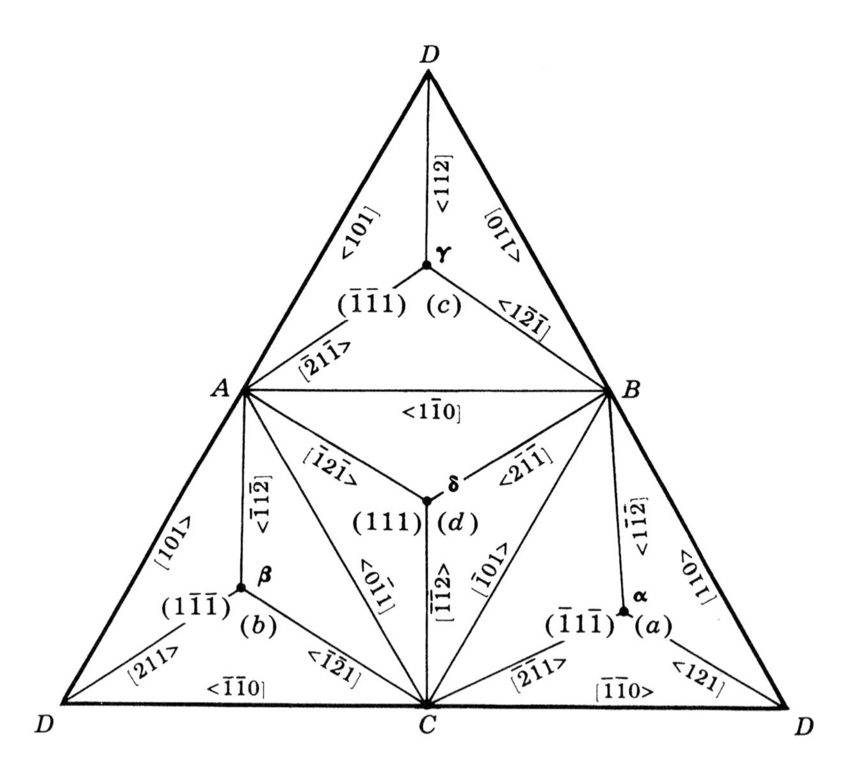

Figure A.1. The Thompson tetrahedron, which is useful for assigning consistent indices. Make an enlarged copy and fold to make your own tetrahedron.

edges correspond to the six $\langle 110 \rangle$ glide directions in the fcc structure. The four apexes are labelled A, B, C, and D, and the mid-points of the faces opposite these apexes are labelled α, β, γ, and δ respectively. A Thompson tetrahedron opened up at the corner D is shown in the figure, with the Miller indices of the planes and directions marked. ABC are ordered in a clockwise sequence, following the original notation of Thompson, and as used, for example, by Hirth and Lothe (1968).

References

Allen C W, Birtcher R C, Donnelley S E, Furuya K, Ishikawa N and Song M 1999 *Appl. Phys. Lett.* **74** 2611

Allen C W, Ohnuki S and Takahashi H 1994 *Trans. Mater. Res. Soc. Jpn.* **17** 93

Allen C W and Ryan E A 1998 *Microsc. Res. Technique* **42** 255–9

Arnold G W 1985 *Scientific Basis for Nuclear Waste Management (Mater. Res. Soc. Symp. Proc., 44)* vol VIII, ed C M Jantzen, J A Stone and R C Ewing (Pennsylvania, PA: Materials Research Society) p 617

——1986 *Radiat. Eff.* **98** 55

Ashby M F and Brown L M 1963 *Phil. Mag.* **8** 1083

Averback R S and Ghaly M 1994 *J. Appl. Phys.* **76** 3908

Bacon D J, Calder A F, Gao F, Kapinos V G and Wooding S J 1995 *Nucl. Instrum. Methods* B **102** 51

Bench M W, Robertson I M and Kirk M A 1991 *Nucl. Instrum. Methods* B **59–60** 372

Bench M W, Robertson I M, Kirk M A and Jenčič I 2000 *J. Appl. Phys.* **87** 49

Bench M W, Tappin D K and Robertson I M 1992 *Phil. Mag. Lett.* **66** 39

Bentley J, Alexander K B and Wang Z L 1990 *Proc. 12th Int. Congress for Electron Microscopy* vol 4, ed L D Peachey and D B Williams (San Francisco, CA: San Francisco Press) p 400

Biersack J P and Haggmark L G 1980 *Nucl. Instrum. Methods* **174** 257

Birtcher R C and Donnelley S E 1996 *Phys. Rev. Lett.* **77** 4374

——1998 *Mater. Chem. Phys.* **54** 111

Birtcher R C, Donnelley S E, Song M, Furuya K, Mitsuishi K and Allen C W 1999 *Phys. Rev. Lett.* **83** 1617

Black T J, Jenkins M L, English C A and Kirk M A 1987 *Proc. R. Soc.* A **409** 177

Brown C 1976 *Proc. EMAG 75* ed J A Venables (New York: Academic) p 405

Bullough R and Newman R C 1960 *Phil. Mag.* **5** 921

Bullough T J, English C A and Eyre B L 1991 *Proc. R. Soc.* A **435** 85

Carpenter G C J 1976 *Phys. Status Solidi* a **37** K61

Castro-Fernandez F R, Sellars C M and Whiteman J A 1985 *Phil. Mag.* A **52** 289

Cawthorne C and Fulton E J 1966 Consultants symposium on the nature of small defect clusters *Harwell Report* AERE-R5269, p 446

Cherns D, Hirsch P B and Saka H 1980 *Proc. R. Soc.* A **371** 213

Chik K P, Wilkens M and Rühle M 1967 *Phys. Status Solidi* **23** 113

Cockayne D J H 1973 *J. Microscopy* **98** 116

Cockayne D J H, Ray I L F and Whelan M J 1969 *Phil. Mag.* **20** 1265

Coene W, Bender H and Amelinckx S 1985 *Phil. Mag.* A **52** 369

Daulton T L, Kirk M A and Rehn L E 2000 *Phil. Mag.* A **80** 809–42
De la Rubia T D, Averback R S, Benedek R and King W E 1987 *Phys. Rev. Lett.* **59** 1930
De la Rubia T D and Guinan M W 1991 *Phys. Rev. Lett.* **66** 2766
DeNatale J F and Howitt D G 1982 *Am. Ceram. Soc. Bull.* **61** 582
——1985 *Radiat. Eff.* **91** 89
——1987 *Am. Ceram. Soc. Bull.* **66** 1393
Devine R A B 1994 *Nucl. Instrum. Methods* B **91** 378
Diepers H and Diehl J 1966 *Phys. Status Solidi* **19** K109
Donnelley S E and Birtcher R C 1997 *Phys. Rev.* B **56** 13 599
Eades J A 1996 *Microscopy Today* **96-1** 24
Edmondson B and Williamson G K 1964 *Phil. Mag.* **9** 277
EerNisse E P 1974 *J. Appl. Phys.* **45** 167
English C A, Eyre B L, Bartlett A F and Wadley H N G 1977 *Phil. Mag.* **35** 533
English C A, Eyre B L and Holmes S M 1980 *J. Phys. F: Met. Phys.* **10** 1065
English C A, Eyre B L and Summers J 1976 *Phil. Mag.* **34** 603
English C A and Jenkins M L 1981 *J. Nucl. Mater.* **96** 341
——1987 *Mater. Sci. Forum* **16–18** 1003
English C A, Phythian W J and Foreman A J E 1990 *J. Nucl. Mater.* **174** 135
Evans J H 1971 *Radiat. Eff.* **10** 55
Ewing R C, Weber W J and Clinard F W Jr 1995 *Prog. Nucl. Energy* **29** 63
Eyre B L 1973 *J. Phys. F: Met. Phys.* **3** 422
Eyre B L and Bullough R 1965 *Phil. Mag.* **12** 31
Eyre B L, Mayer D M and Perrin R C 1977a *J. Phys. F: Met. Phys.* **7** 1359
——1977b *J. Phys. F: Met. Phys.* **7** 1370
Föll H and Wilkens M 1975 *Phys. Status Solidi* a **31** 519
Foreman A, Phythian W J and English C A 1992 *Phil. Mag.* **66** 671
Foreman A, von Harrach H S and Saldin D 1980 *Inst. Phys. Conf. Ser.* **52** 225
——1982 *Phil. Mag.* A **45** 625
Frischherz M C, Kirk M A, Zhang J P and Weber H W 1993 *Phil. Mag.* A **67** 1347
Fukushima H, Jenkins M L and Kirk M A 1997a *Phil. Mag.* A **75** 1567
——1997b *Phil. Mag.* A **75** 1583
Fukushima H, Shimomura Y, Guinan M W and Kiritani M 1989 *Phil. Mag.* A **60** 415
Fukushima H, Shimomura Y, Kiritani M, Guinan M W, Gerstenberg H, Mukouda I and Mitoma T 1994 *J. Nucl. Mater.* **212–215** 154
Fukushima H, Shimomura Y and Yoshida H 1991 *J. Nucl. Mater.* **179–181** 939
Furuya K, Ishikawa N and Allen C W 1999 *J. Microscopy* **194** 152
Giapintzakis J, Ginsberg D M, Kirk M A and Ockers S 1994 *Phys. Rev.* B **50** 15 967
Giapintzakis J, Lee W C, Rice J P, Ginsberg D M, Robertson I M, Wheeler R, Kirk M A and Ruault M-O 1992 *Phys. Rev.* B **45** 10 677
Gibson J M and Treacy M M J 1997 *Phys. Rev. Lett.* **78** 1074
Goodhew P J and Humphreys F J 1988 *Electron Microscopy and Analysis* 2nd edn (London: Taylor and Francis) ISBN 0-85066-415-2, ISBN 0-85066-414-4(pbk)
Griffiths M, Gilbon D, Regnard C and Lemaignan C 1993 *J. Nucl. Mater.* **205** 273
Groves G W and Kelly A 1961 *Phil. Mag.* **6** 1527
——1962 *Phil. Mag.* **7** 892
Grüschel W 1979 *Dr.rer.nat Thesis* University of Stuttgart
Habibi-Bajguirani H R and Jenkins M L 1996 *Phil. Mag. Lett.* **73** 155
Hardouin-Duparc H, Doole R C, Jenkins M L and Barbu A 1996 *Phil. Mag. Lett.* **71** 325

Hardy G J and Jenkins M L 1985 *Phil. Mag.* A **52** L19
Harry T and Bacon D J 1997 *Mater. Res. Soc. Symp. Proc.* **439** 495
Hashimoto H, Howie A and Whelan M J 1962 *Proc. R. Soc.* A **269** 80
Häussermann F 1972 *Phil. Mag.* **25** 561
Häussermann F, Katerbau K-H, Rühle M and Wilkens M 1973 *J. Microscopy* **98** 135
Häussermann F, Rühle M and Wilkens M 1972 *Phys. Status Solidi* b **50** 445
Hertel B 1979 *Phil. Mag.* A **40** 313 and 331
Hertel B and Diehl J 1976 *Phys. Status Solidi* a **33** K73
Hertel B, Diehl J, Gotthardt R and Sultze M 1974 *Applications of Ion Beams to Metals* ed S T Picraux, E P EerNisse and F L Vrok (New York: Plenum) p 507
Hertel B and Rühle M 1979 *Proc. Conf. on Irradiation Behaviour of Metallic Materials for Fast Reactor Core Components (Corsica)* p 219
Hirsch P B 1978 *Electron Diffraction 1927–1977* ed P J Dobson, J B Pendry and C J Humphreys (London: Institute of Physcis) p 335
——1991 *Ultramicroscopy* **37** 1
Hirsch P B, Howie A, Nicholson R B, Pashley D W and Whelan M J 1977 *Electron Microscopy of Thin Crystals* (Malabar, FL: Krieger) ISBN 0-88275-376-2 revised edn, first published 1965, second revision 1977
Hirth J P and Lothe J 1968 *Theory of Dislocations* (New York: McGraw-Hill)
Hobbs L W 1994 *Nucl. Instrum. Methods* B **91** 30
Hoelzer T and Ebrahimi F 1995 *Mater. Res. Soc. Symp. Proc.* **373** 57
Holmes S M, Eyre B L, English C A and Perrin R C 1979 *J. Phys. F: Met. Phys.* **9** 2307
Horton L L, Bentley J and Farrell K 1982 *J. Nucl. Mater.* **108/109** 22
Howe L M 1994 *Intermetallic Compounds: Vol. 1, Principles* ed J H Westbrook and R L Fleisher (Chichester: Wiley) ch 34
Howe L M and Rainville M H 1977 *J. Nucl. Mater.* **68** 215
——1991 *Surface Chemistry and Beam-Solid Interactions (Mater. Res. Soc. Symp. Proc., 201)* (Pennsylvania, PA: Materials Research Society) p 215
Howe L M, Rainville M H, Haugen H K and Thompson D A 1980 *Nucl. Instrum. Methods* **170** 419
Howie A and Basinski Z S 1968 *Phil. Mag.* **17** 1039
Howie A and Whelan M J 1961 *Proc. R. Soc.* A **263** 217
Ipohorski M and Brown L M 1970 *Phil. Mag.* **22** 931
Ishino S, Sekimura N, Hirooka K and Muroga T 1986 *J. Nucl. Mater.* **141–143** 776
Jäger W and Wilkens M 1975 *Phys. Status Solidi* a **32** 89
Jenkins M L 1974 *Phil. Mag.* **29** 813
——1994 *J. Nucl. Mater.* **216** 124
Jenkins M L, Cockayne D J H and Whelan M J 1973 *J. Microscopy* **98** 155
Jenkins M L and English C A 1982 *J. Nucl. Mater.* **108–109** 46
Jenkins M L, English C A and Eyre B L 1978 *Phil. Mag.* A **38** 97
Jenkins M L and Hardy G J 1987 *Mater. Sci. Forum* **15–18** 1075
Jenkins M L, Hardy G J and Kirk M A 1987 *Mater. Sci. Forum* **15–18** 901
Jenkins M L, Katerbau K-H and Wilkens M 1976 *Phil. Mag.* A **34** 1141
Jenkins M L, Kirk M A and Fukushima H 1999a *J. Electron Microscopy* **48** 323
Jenkins M L, Kirk M A and Phythian W J 1993 *J. Nucl. Mater.* **205** 16
Jenkins M L, Mavani P, Müller S, and Abromeit C 1999b *Microstructural Processes in Irradiated Metals (Mater. Res. Soc. Symp. Proc., 540)* (Pennsylvania, PA: Materials Research Society) p 515

Jenkins M L, Norton N G and English C A 1979 *Phil. Mag.* A **40** 131
Jenkins M L and Wilkens M 1976 *Phil. Mag.* A **34** 1155
Katerbau K-H 1976 *Phys. Status Solidi* a **38** 463
Kelly P M, Jostens A, Blake R G and Napier J G 1975 *Phys. Status Solidi* a **31** 771
Kenik E A 1994 *J. Nucl. Mater.* **216** 157
King S L, Jenkins M L, Kirk M A and English C A 1992 *Effects of Irradiation on Materials: 15th Int. Symposium (American Society for Testing of Materials, Special Technical Publication 1125)* p 448
——1993 *J. Nucl. Mater.* **205** 467
King W E, Merkle K L and Meshii M 1983 *J. Nucl. Mater.* **117** 12
Kiritani M 1992 *Mater. Sci. Forum* **97–99** 141
——1994 *J. Nucl. Mater.* **216** 220
Kiritani M, Fukuda Y, Mima T, Iiyoshi E, Kizuka Y, Kojima S and Matsunami N 1994 *J. Nucl. Mater.* **212–215** 192
Kiritani M, Yoshiie T, Kojima S and Satoh Y 1993 *J. Nucl. Mater.* **205** 460
Kirk M A, Jenkins M L and Fukushima H 2000 *J. Nucl. Mater.* **276** 50
Kirk M A and Yan Y 1999 *Micron* **30** 507
Kojima S, Satoh Y, Taoka H, Ishida I, Yoshiie T and Kiritani M 1989 *Phil. Mag.* A **59** 519
Krivanek O L, Dellby N and Lupini A R 1999 *Ultramicroscopy* **78** 1
Laupheimer A 1981 *Dr.rer.nat Thesis* University of Stuttgart
Lee W E, Jenkins M L and Pells G P 1985 *Phil. Mag.* A **51** 639
Little E A 1972 *Radiat. Eff.* **16** 135
Little E A, Bullough R and Wood M H 1980 *Proc. R. Soc.* A **372** 565
Loretto M H 1984 *Electron Beam Analysis of Materials* (London: Chapman and Hall) ISBN 0-412-23390-8, ISBN 0-412-3400(pbk)
Maher D M and Eyre B L 1971 *Phil. Mag.* **23** 409
Maiti A, Pantelides S T, Chisholm M F and Pennycook S J 1999 *Appl. Phys. Lett.* **75** 2380
Marples J A C 1988 *Nucl. Instrum. Methods Phys. Res.* B **32** 480
McIntyre K G 1967 *Phil. Mag.* **15** 205
Merkle K L 1966 Consultants symposium on the nature of small defect clusters *Harwell Report* AERE-R5269, p 8
Merkle K L and Jäger W 1981 *Phil. Mag.* A **44** 741
Merkle K L, Singer L R and Hart R K 1963 *J. Appl. Phys.* **34** 2800
Michal G M and Sinclair R 1979 *Phil. Mag.* A **42** 691
Mitchell J B and Bell W L 1976 *Acta. Met.* **24** 147
Monzen R, Jenkins M L and Sutton A P 2000 *Phil. Mag.* A **80** 711
Mooh H W and Kruit P 1999 *Nucl. Instrum. Methods* A **427** 109
Muncie J M, Eyre B L and English C A 1985 *Phil. Mag.* A **52** 309
Muncie J W 1980 *DPhil Thesis* University of Sussex
Newbury D E, Wollman D, Irwin K, Hilton G and Martinis J 1999 *Ultramicroscopy* **78** 73
Nicol A, Jenkins M L and Kirk M A 1999 *Microstructural Processes in Irradiated Metals (Mater. Res. Soc. Symp. Proc., 540)* (Pennsylvania, PA: Materials Research Society) p 409
Ohr S M 1972 *Phil. Mag.* **26** 1307
——1977 *Proc. 35th Annual Meeting of Electron Microscopy Society of America* p 52
——1979 *Phys. Status Solidi* a **56** 527
Okamoto P R, Lam N Q and Rehn L E 1999 *Solid State Phys.* **52** 1
Okamoto P R and Wiedersich H 1974 *J. Nucl. Mater.* **53** 336

Othen P J, Jenkins M L and Smith G D W 1994 *Phil. Mag.* A **70** 1
Othen P J, Jenkins M L, Smith G D W and Phythian W J 1991 *Phil. Mag.* A **64** 383
Pennycook S J 1992 *Microscopy: The Key Research Tool* (Woods Hole, MA: The Electron Microscopy Society of America) p 51
Pennycook S J and Jesson D E 1991 *Ultramicroscopy* **37** 14
Phythian W J, Eyre B L, Bacon D J and English C A 1987 *Phil. Mag.* A **55** 757
Piskunov D I 1985 *Sov. Phys. Solid State* **27** 2125
Primak W, Fuchs L H and Day P 1953 *Phys. Rev.* **92** 1064
Robertson I M 1982 *DPhil Thesis* University of Oxford.
Robertson I M and Jenčič I 1996 *J. Nucl. Mater.* **239** 273
Robertson I M, Jenkins M L and English C A 1982 *J. Nucl. Mater.* **108/109** 209
Robertson I M, Vetrano J S, Kirk M A and Jenkins M L 1991 *Phil. Mag.* A **63** 299
Robinson T M and Jenkins M L 1981 *Phil. Mag.* A **43** 999
Ruault M O, Bernas H and Chaumont J 1979 *Phil. Mag.* A **39** 757
Rühle M 1967 *Phys. Status Solidi* **19** 263
Rühle M 1972 *Radiation Induced Voids in Metals* ed J W Corbett and L C Iannello (US Atomic Energy Commission) CONF-710601
Rühle M, Häussermann F and Rapp M 1970 *Phys. Status Solidi* **39** 609
Rühle M and Wilkens M 1975 *Cryst. Lattice Defects* **6** 129
Rühle M, Wilkens M and Essman U 1965 *Phys. Status Solidi* **11** 819
Sagaradze V V, Lapin S S, Kirk M A and Goshchitskii B N 1999 *J. Nucl. Mater.* **274** 287
Saldin D K, Stathopoulos A Y and Whelan M J 1979 *Phil. Trans. R. Soc.* **292** 524
Saldin D K and von Harrach H 1980 *Inst. Phys. Conf. Ser.* vol 52, ed T Mulvey, p 229
Sass S L and Eyre B L 1973 *Phil. Mag.* **27** 1447
Satoh Y, Taoka H, Kojima S, Yoshiie T and Kiritani M 1994 *Phil. Mag.* A **70** 869
Schäublin R, Dai Y, Osetsky Y and Victoria M 1998 *'Electron Microscopy 1998' Proc. ICEM14 (Cancun, Mexico)* Symposium C, vol 1 (Bristol: Institute of Physics) p 173
Shepherd B W O 1981 *DPhil Thesis* University of Oxford
Shepherd B W O, Jenkins M L and English C A 1987 *Phil. Mag.* A **56** 458
Shepherd C M 1990 *J. Nucl. Mater.* **175** 170
Silcox J and Hirsch P B 1959 *Phil. Mag.* **4** 1376
Sigle W, Jenkins M L and Hutchison J L 1988 *Phil. Mag. Lett.* **57** 267
Sinclair R, Michal G M and Yamashita T 1981 *Metall. Trans.* A **12** 1503
Spence J C H 1988 *Experimental High-Resolution Electron Microscopy* 2nd edn (Oxford: Oxford University Press) ISBN 0-19-505405-9
Stathopoulos A Y 1977 *DPhil Thesis* University of Oxford
——1981 *Phil. Mag.* A **44** 285
Stadelmann P A 1987 *Ultramicroscopy* **21** 131
Steeds J W and Mansfield J 1984 *Convergent Beam Analysis of Alloy Phases* (Bristol: Hilger) ISBN 0-85274-771-3
Stoneham A M 1994 *Nucl. Instrum. Methods Phys. Res.* B **91** 1
Storey B G, Kirk M A, Osborne J A, Marks L D, Kostic P and Veal B W 1996 *Phil. Mag.* A **74** 617
Sykes L J, Cooper W D and Hren J J 1981 *Oak Ridge Report* ORNL/TM-7619
Takeyama T, Ohnuki S and Takahashi H 1981 *Trans. Iron Steel Inst. Japan* **21** 326
Tanaka K, Kimoto K and Mihama K 1991 *Ultramicroscopy* **39** 395
Thompson N 1953 *Proc. Phys. Soc.* B **66** 481
Titchmarsh J M and Dumbill S 1996 *J. Nucl. Mater.* **227** 203

——1997 *J. Microscopy* **188** 224
Titchmarsh J M and Williams T M 1981 *Proc. Conf. on Quantitative Microanalysis with High Spatial Resolution* Book 277 (London: The Metals Society)
Treacy M M J and Gibson J M 1996 *Acta Crystallogr.* A **52** 212
Urban K and Yoshida N 1981 *Phil. Mag.* A **44** 1193
Vetrano J S, Robertson I M, Averback R S and Kirk M A 1992 *Effects of Irradiation on Materials: 15th Int. Symposium (American Society for Testing of Metals, Special Technical Publication 1125)* p 375
Vetrano J S, Robertson I M and Kirk M A 1990 *Scripta Metallurgica Materialia* **24** 157
von Harrach H and Foreman A J E 1980 *Radiat. Eff.* **46** 7
Walker R S and Thompson D A 1978 *Radiat. Eff.* **37** 113
Wilkens M 1964 *Phys. Status Solidi* **6** 939
——1966 *Phys. Status Solidi* **13** 529
——1978 *Diffraction and Imaging Techniques in Materials Science* ed S Amelinckx, R Gevers and J van Landuyt (Amsterdam: North-Holland) p 185
Wilkens M, Jenkins M L and Katerbau K-H 1977 *Phys. Status Solidi* a **39** 103
Wilkens M, Rühle M and Häussermann F 1973 *J. Microscopie* **16** 199
Williams D B 1983 *Practical Analytical Electron Microscopy in Materials Science* Philips Electron Inst. Inc., Electron Optics Publishing Group, ISBN 0-9612934-0-3
Williams D B and Carter C B 1996 *Transmission Electron Microscopy* (Oxford: Plenum) ISBN 0-306-45324-X (pbk)
Williams T M and Eyre B L 1976 *J. Nucl. Mater.* **59** 18
Williams T M, Titchmarsh J M and Arkell D R 1982 *J. Nucl. Mater.* **107** 222
Wilson M M 1971 *Phil. Mag.* **24** 1023
Wilson M M and Hirsch P B 1972 *Phil. Mag.* **25** 983
Yan Y and Kirk M A 1998 *Phys. Rev.* B **57** 6152
Yoshida N and Urban K 1980 *Phil. Mag. Lett.* **75** 231
Zhu Y, Budhani R C, Cai Z X, Welch D O, Suenaga M, Yoshizaki R and Ikeda H 1993 *Phil. Mag. Lett.* **67** 125

Index